Vesalius: The China Root Epistle

ℰ A NEW TRANSLATION AND CRITICAL EDITION

This book provides the first annotated English translation from the original Latin of Andreas Vesalius' *The China Root Epistle*. Ostensibly his appraisal of a fashionable herbal remedy, *The China Root Epistle* concentrates on Vesalius' skeptical appraisal of traditional Galenic anatomy, which was based on animal rather than human dissections. Along with reflections about his life as a young anatomist, Vesalius argued that the new science of anatomy should devote itself less to rhetorical polemics and more to the craft of direct observation based on human dissection. This volume provides annotations to link the Epistle with Vesalius' earlier and more famous *On the Fabric of the Human Body*, and includes illustrations from the famous woodcuts first used in the 1543 edition of the *Fabrica*.

Daniel H. Garrison is Professor Emeritus in the Department of Classics at Northwestern University. He is the translator of *The Fabric of the Human Body* (with Malcolm Hast) and the author of several books, including *Sexual Culture in Ancient Greece*, *The Student's Catullus*, and *Horace Epodes and Odes: A New Annotated Latin Edition*.

Vesalius:
The China Root Epistle

A New Translation and Critical Edition

ANDREAS VESALIUS

EDITED AND TRANSLATED BY
DANIEL H. GARRISON
Northwestern University

WITH ADDED ILLUSTRATIONS FROM THE
1543 AND 1555 *De humani corporis fabrica*

CAMBRIDGE
UNIVERSITY PRESS

CAMBRIDGE
UNIVERSITY PRESS

Shaftesbury Road, Cambridge CB2 8EA, United Kingdom

One Liberty Plaza, 20th Floor, New York, NY 10006, USA

477 Williamstown Road, Port Melbourne, VIC 3207, Australia

314–321, 3rd Floor, Plot 3, Splendor Forum, Jasola District Centre, New Delhi – 110025, India

103 Penang Road, #05–06/07, Visioncrest Commercial, Singapore 238467

Cambridge University Press is part of Cambridge University Press & Assessment,
a department of the University of Cambridge.

We share the University's mission to contribute to society through the pursuit of
education, learning and research at the highest international levels of excellence.

www.cambridge.org
Information on this title: www.cambridge.org/9781107026353

First published 2015

A catalogue record for this publication is available from the British Library

Library of Congress Cataloging-in-Publication data
Vesalius, Andreas, 1514–1564.
[Epistola, rationem modumque propinandi radicis Chynae decocti. English]
Vesalius, the China root epistle : a new translation and critical edition /
Andreas Vesalius; [translated and edited by] Daniel H. Garrison;
with added illustrations from the 1543 and 1555 De humani corporis fabrica.
    p. ; cm.
Includes bibliographical references and index.
ISBN 978-1-107-02635-3 (hardback)
I. Garrision, Daniel H. II. Title.
[DNLM: 1. Galen. 2. Dubois, Jacques, 1478–1555. 3. Anatomy. 4. Phytotherapy.
5. Plant Preparations – therapeutic use. 6. Smilax. WZ 290]
611–dc23   2012008272

ISBN   978-1-107-02635-3   Hardback

ℰ⌢ CONTENTS

(Page numbers of the 1546 edition appear in italic.
Bracketed heads do not appear in the 1546 edition.)

When Vesalius' friend Bernardo Navagero, the Venetian ambassador to the court of Charles V, fell ill at Nymwegen in the Netherlands and was not well enough to travel, Charles V assigned Vesalius the task of staying behind to care for him. It was early January 1546, three years since the publication of his epochal atlas of human anatomy, *De humani corporis fabrica*. For about twelve weeks, with little else to do, Vesalius thought and wrote about his work as an anatomist and its meaning for the discipline. The result of his reflection was the book here translated. Its significance lies to a great degree in what Vesalius had to say about the method that would eventually become what we call scientific.

For most of its pages, the *China Root Epistle* is Vesalius' sometimes barbed response to Jacobus Sylvius' vendetta which later (1551) came to print as *Vaesani cujusdam calumniarum in Hippocratis Galenique rem anatomicam depulsio* (A Refutation of Calumnies by a Certain Madman against Hippocratic and Galenic Anatomy), which maliciously turned the genitive *Vesalii* into *Vaesani* "madman."

What was the cause of Sylvius' malice? When Vesalius was a medical student in Paris from 1533 to 1536 he had been (to judge from his earliest remarks in print) an ardent disciple of Sylvius, who in defiance of tradition performed his own dissections while he lectured. Vesalius had probably earned his teacher's favorable attention, though nothing is recorded about the actual relationship between the two. Sylvius (Jacques Dubois) was a committed humanist who believed that the Ancients wrote nothing wrong, and that the best their latter-day admirers could do was to transmit ancient Greek (as opposed to medieval Arabic) wisdom uncontaminated. In taking this position, Sylvius and his fellow humanists placed their faith in personalities, especially that of Galen of Pergamon (AD 129–199 or 216), rather than a method.

When Vesalius moved on to Padua in 1537, he began to make a reputation there and at Bologna pointing out Galen's errors that resulted from projecting animal anatomy onto humans. Sylvius would have seen this as a treacherous abandonment of the humanist faith. When Sylvius wrote about anatomy, he skirted the errors of Galen in silence. Vesalius' repeated, insistent, and overt assertion of Galen's errors would have seemed flamboyant and insolent to Sylvius' cautious but caustic nature. To the mind of Sylvius, the controversy about anatomy had been poisoned by disloyalty. Worse yet, it had become clear soon after 1543 that Vesalius' anti-Galenic *Fabrica* was destined to eclipse any Galenic anatomy book Sylvius could aspire to write.

Vesalius' response in the *China Root Epistle* to Sylvius' attacks was in the first place a fresh articulation of Galen's many errors and in the second an effort to protest that he was no traitor to the humanist cause. The truth was more complicated. A great deal of what Vesalius had published three years earlier about the fabric of the human body was still Galenic and left many of Galen's errors and other faults of traditional anatomy unchallenged. At the same time, even while he was writing the *Fabrica* Vesalius' own thinking had evolved and he could no longer be the faithful disciple of the Ancients, as he wished to be perceived. To fit that paradox, Vesalius needed to throw some rhetorical sand in his readers' eyes.

To start with, he draped his response to Sylvius in the sheep's clothing of a monograph on a fashionable herbal remedy that was in great demand by wealthy patients who suffered from any of the three great scourges of the age: syphilis, gout, and stone. All three were practically incurable, and seemed to concentrate their attacks upon the most successful and gifted: stone, for example, afflicted Thomas Linacre (1460–1524), Desiderius Erasmus (1466–1536), Francis Walsingham (1532–1590), Michel de Montaigne (1533–1592), Oliver Cromwell (1599–1658), John Dryden (1631–1700), Samuel Pepys (1633–1703), and Isaac Newton (1643–1727).

Gout, "the patrician malady," tortured the Medici patriarch Cosimo de' Medici (1389–1464), his son Piero il Gottoso "the Gouty" (1416–1469), Vesalius' patron Charles V, his medical colleague Ambroise Paré (1510–1590), Elizabeth I's chief advisor William Cecil (1521–1598), and scores of other notables. Because syphilis, the *mal Francese*, bore the stigma of sexual incontinence, its victims were often unacknowledged. They may have included Cesare Borgia (1475–1507), the English monarch Henry VIII (1491–1547), and the Russian Czar Ivan the Terrible (1530–1584), and certainly included numerous princes of the Church, including Giuliano della Rovere (1443–1513), who became Pope Julius II in 1503. Its later victims included Franz Schubert, who died at 31 in 1828.

The demand for a cure was insistent and well funded, as was the demand for professional appraisals of the most celebrated treatments for all three afflictions. Such an appraisal by the author of the *Fabrica*, who was also a member of the Holy Roman Emperor's personal medical staff, was sure to circulate widely in the medical community. It was published twice (Basel and Venice) in 1546, the year of its completion, and a third time the next year in Lyon. A German version dealing only with the China root question was published in Würzburg in 1548. More Latin reprints appeared in 1566, 1599, and 1728.[1]

Though not the main subject of the monograph that bears its name, the China root was a hot topic in 1546. The eponymous herbal was the rootstock of *smilax china*, a plant native to the East Indies that is still used in traditional Chinese medicine. Its aqueous extract is believed to have anti-inflammatory and analgesic properties,[2] making it roughly comparable to aspirin (whatever additional placebo effects it may have had). Introduced into Europe as early as 1525 and a widely known specific against syphilis by 1535,[3] it was thought to promote perspiration and urine and was used for a variety of other diseases, including gout and stone. It remained unmentioned by the leading botanists of the time.[4] But by 1546, it was a celebrated panacea demanded by wealthy clients such as Vesalius' own patron, the Holy Roman Emperor Charles V. Vesalius' professional judgment of the China root was at best polite and agnostic, but long before the end of his *Epistle* he dismissed it as *stupidus* and devoted the rest of his monograph to topics that were closer to his heart: his insistence that Galen was an unreliable authority on human anatomy because he had dissected animals instead of humans, and his defense against Sylvius' scurrilous attacks.

It is likely that some of this larger part of the *Epistle* originated from the projected annotations to the anatomical works of Galen that

---

[1] Cushing 1962, 163–7.

[2] Shu et al. 2006. The rhizome of a related plant, *Smilax glabra*, known in English as the glabrous greenbrier rhizome, has been used recently in combination with other Chinese herbs in the treatment of syphilis (Bensky & Gamble 1986, 144f.).

[3] Schmitz and Tan 1967, 221.

[4] For example, Johannes Ruellius *De medicamentorum compositione* (1540), Leonhard Fuchs *De historia stirpium* (1542), Pierandrea Mattioli *Commentarii in sex libros Pedacii Dioscoridis* (1544), Johannes Actuarius *De medicamentorum compositione* (1546).

Vesalius referred to at the end of his chapter in the *Fabrica* on the flexors and extensors of the radius, Bk. II ch. 46: "I shall reveal all of this in my annotations to the Anatomical Works of Galen, which I have already well begun and shall at some time publish separately or together with the books of Galen much better corrected than formerly."[5] But near the end of this *Epistle* (page 195 of the Basel edition) Vesalius says that he burned those annotations along with other writings when he left his academic post at Padua to enter the service of Charles V. By that time, he says, his notes on Galen had grown into a massive work, *ingens volumen*. Either he reconstructed some of them from memory for the *Epistle* or he exaggerated when he said he had burned them. Whatever the case, the length of his remarks in the *Epistle* on the errors of Galen reflects the regret he expresses here at the *petulentia* with which he had abandoned his research as an anatomist.

Taken in this way, the *Epistle* may be thought of as the core of the lost commentary on Galen, drawing upon what Vesalius had already published in the *Fabrica* and expanding those remarks into a sustained polemic using a series of discrepancies between Galen's animal-based anatomy and Vesalius' anatomy founded upon human dissection. It testifies more than the *Fabrica* to Vesalius' substantial work in comparative anatomy, particularly in parallel dissections of human parts with corresponding parts of common mammals and caudate and non-caudate simians, with a view to identifying which parts corresponded to Galen's descriptions and which did not. The *Epistle* reminded its readers that the case against Galenism was massive, pervading the functions as well as the fabric of the entire human body.

The rapidity with which the *Epistle* was written and the lack of an editor resulted in some repetitiveness in the later pages, where

---

[5] On these abandoned projects, see O'Malley 1964, 190 f., 223. Nancy Siraisi has called this type of commentary, in which Giovanni Argenterio also engaged, a "counter-commentary" because it concentrated on pointing out errors. (Siraisi 1990, 172).

Vesalius restates criticisms of Galen he had mentioned earlier. Though these repetitions give those pages a rambling quality, they also reveal some errors of Galen that were particularly on the author's mind. These include the observation that the omohyoid muscle does not move the scapula (pp. 58, 159), the correct length of the styloid process of the ulna (pp. 104, 154), and the function of the pancreas relative to the lower orifice of the stomach (pp. 111, 172). The most persistent repetition regards the absence of any vessel that might convey black bile from the spleen to the stomach (pp. 133, 135, 138, 173). Though Vesalius refrains from mentioning the effect of this vascular hiatus upon the folklore of melancholia and humoral medicine generally, it may be speculated that the question was something he hoped his readers would take up.

In centering his *Epistle* on criticisms of Galen and Galenism, Vesalius wished to avoid being perceived as a traitor to the humanist cause, which aimed to restore the pristine dominance of Greek medicine, still held up as the one and only *prisca medicina*. He sometimes, therefore, casts Galen as the wrong-headed detractor of the Ancients who substituted animal anatomy for the human anatomy in which they (especially the Alexandrians of the 3rd century BC) were supposedly versed. Just before the middle of the *Epistle* he stakes out his ground as the champion of Greek (as opposed to Galenic) anatomy:

> When Galen cut up his monkeys and saw that they differed from the description of the ancients, who trained themselves on human dissections, he did not scruple to state that they had not seen that [third] fiber of the lung and who knows what else. I should therefore be thought more impious if I had not vindicated those Ancients with a true description of the human fabric. If because of the powerful devotion to Galen under which I labor and my special regard for him I were to leave his opinions everywhere undisturbed contrary to the testimony of my eyes

and the truth of the matter, I should be willing to have my generation wander in confusion like all the ages that have followed Galen, and let his misrepresentation of the Greeks go undetected.[6]

As he draws nearer the end of the *Epistle*, Vesalius articulates a critique of medical work that sets aside the vendettas and polemics with which Galen and his successors (including Vesalius himself) had overly occupied themselves. The craft of medicine, he says, is not about criticizing the books others have written or the authors who wrote them but about

> the diligent and careful dissection of humans, simians, and certain other animals. Nor is it sufficient to occupy oneself in speaking ill of someone or ridiculing the efforts of others and to detract equally from one's own and others' glory ... when one should rather be working up a sweat in common efforts at the truth, and believing that we too were born human. Something in the vast art of medicine may be present in us, as well as a faculty of discovery, if we are more strongly held by a desire for truth than for calumniating others.

In making this remark he is not only distancing himself from the critiques of Galen and Galenists to which he had devoted his writing career, he is also distancing medical research from personal rivalry and from the philological work that humanist scholars like Cornarius were doing, and asserting its focus upon "diligent and careful dissection."

---

[6] Galenus enim, quum simias suas scinderet, illasque a veterum qui hominum dissectionibus sese exercebant historia abesse videret, non veritus est testari, eos illam pulmonis fibram, & nescio quae alia, latuisse. Adeo ut magis impius censeri deberem, si in vera hominis fabricae historia Veteres illos non excusassem: quam si propter insignem quo erga Galenum laboro affectum, singularemque observantiam, illius placita undique imperturbata reliquissem: atque hoc nostrum seculum, perinde atque omnia quae Galenum secuta sunt, hallucinari, & Graecorum imposturam latere voluissem. (p. 95).

Vesalius is no longer thinking like a humanist; he is and beginning to think like a scientist.

It is clear from a letter that Vesalius sent to Thomas Gast, his friend in Basel, that he set great store by this little publication:

> I should be happy to have the work published soon and in elegant format. I request you to advise Oporinus to use the best paper and to see that the book has wide margins. I shall bear the extra cost. Thereby the printing is clearer and the work of the typographer made easier. The larger a book is, the greater my pleasure in it. I know you will laugh at my wishes; nevertheless, I wish it. Nothing gives me more pleasure than a splendid edition of my work. ... Impress it upon Oporinus that he is not, as is his custom, to allow my manuscript to remain for a long time in his drawer.[7]

VESALIUS ALWAYS WROTE BEST WHEN DESCRIBING A PROCEDURE such as a dissection (in the *Fabrica*) or the preparation of a decoction (in the *Epistle*). His language is most tortured in his polemical mode, when he is putting his mental agility on display. Vesalius took from Galen and from the worst vices of medieval scholasticism a preference for dense polemic where we should have preferred clear exposition and economy of language.

His Latinity, seldom transparent at any time, is especially slap-dash in the *Epistle*, changing constructions in mid-sentence, moving in weird ellipses, using nonstandard constructions, and more than once abandoning the rules of syntax altogether. This could be the result of lacking an editor for the *Epistle* or of writing even more hastily than he had for the 1543 *Fabrica*. It could also be the result of

---

[7] English translation from O'Malley 1964, 455 n. 149.

the transmission of his text, which appears from the preface written by Vesalius' brother Franciscus to be a fourth-generation copy. From the autograph (first generation), Jacob Scepper made a copy (second generation) to carry to Ferrara; a copy of that was then made (third generation, by Franciscus) for delivery to Vesalius' publisher Oporinus in Basel, whose printed *Epistle* became the fourth generation. O'Malley speculates that Vesalius sent a revised version of his monograph to Oporinus containing his corrections of the copy set to type by the printer,[8] but we see little evidence of a careful recension by the author before it went to press. Whatever the case, this translation is based upon Oporinus' *editio princeps*. The Latin, sometimes rapid and clear, more often falls into a congested state that requires careful unpacking and diligent guesswork.

Vesalius is not the only important author whose prose was notoriously unreadable. Writing in the first century BC, the literary critic and historian Dionysius of Halicarnassus said about Thucydides' Greek "If people actually spoke like this, not even their mothers or their fathers would be able to tolerate the unpleasantness of it; in fact they would need translators, as if they were writing in a foreign language."[9] Vesalius' Latin in the *Epistle* makes it a foreign language even to the lifelong reader of Latin. It shows an impatience with the language of the humanists which he increasingly seems to have felt had become an end in itself rather than the means to an end. Yet instead of trying to make it more transparent, he made it still more opaque.

Vesalius' language is insistently visual and his working vocabulary is immense.[10] Though humanist Latin endeavored to use only

---

[8] O'Malley 1964, 455 n. 149.
[9] Quoted by Mary Beard in "Which Thucydides Can You Trust?" *New York Review of Books* LVII.14 (September 30, 2010), p. 52.
[10] Like every serious writer of his time in Europe, Vesalius was influenced by Erasmus' *De copia*, first sketched in 1499 but emended and expanded throughout his life until Froben's edition of 1534, which begins "The speech of man is a magnificent and impressive thing when it surges along like a golden river, with thoughts and words pouring out in rich abundance." (tr. Betty I. Knott, *De Copia. Foundations of the*

the classical vocabulary, this would have been difficult in a field such as medicine which had for centuries been developing new words and meanings. As we have already seen, it is no longer strictly accurate to call the Vesalius of 1546 a humanist.

The resulting complexity of Vesalius' Latin makes a literal translation more difficult than it would be for a classical author (not including Thucydides) who was a native speaker. Some word meanings have to be backed out of the Oxford English Dictionary, and many eccentric constructions can only be paraphrased. As when translating the *Fabrica*, I have not tried to mask the way Vesalius wrote, though I have always tried to make it clear and I regularly break down sentences that ramble on too long. The resulting English will not resemble the crisp, efficient language we have been taught to write, because the canons of Early Modern style favored bulk and complexity over concision. What Vesalius feared most to write was something that would seem *sterilis* or *ieiunus*, barren or meager.[11]

In preparing this annotated translation I have mapped the most important links to the 1543 *Fabrica* which serve as a background to what Vesalius wrote in the *Epistle*. Of course, the *Epistle* is not simply a recitation or précis of what was in the *Fabrica*, being often more detailed in its account of Galen's errors and sometimes offering new evidence that Galen did not dissect human cadavers. The annotations here sometimes repeat what I wrote for the *Fabrica* where that is relevant, and where it might be helpful I supply the *nomina anatomica* that my co-author Malcolm Hast provided for the *Fabrica*, with his kind permission.

Except for the portrait of Vesalius in the frontispiece, three large historiated capitals, two of a smaller type, and a colophon figure of Arion, the *Epistle* was printed without illustrations. Partly to relieve

---

Abundant Style. Vol. 24, Collected Works of Erasmus. Toronto: University of Toronto Press, 1974).

11 *ut non quantum sterili mea ieiunaque dictione datum fuit* (p. 44 in the 1546 *Epistola*).

the tedium of two hundred pages of unbroken text without paragraph or chapter breaks, and more importantly to help the reader visualize what Vesalius saw in the human body, I have interpolated several illustrations from the *Fabrica*. Some of these are details of larger woodcuts, and most have been reduced in scale lest they overwhelm the smaller page size of the *Epistle*.

For an online scanned facsimile of the *editio princeps* of the *Epistle* I am grateful to the Bibliothèque interuniversitaire Santé (or Biusanté) at the Université Paris Descartes, and to Northwestern Library's bibliographer William A. McHugh for locating this PDF. To Karen Reeds and an anonymous second referee for the Press, I am grateful for numerous corrections and suggested additions to the notes. The result has been a clearer window into the world that the *China Root Epistle* reveals.

Text

(Page breaks of the 1546 edition are indicated by an oversize asterisk.
The corresponding page number of the 1546 edition appears in the margin.)

ANDREAS VESALIUS OF BRUSSELS
Imperial Physician

EPISTLE

Explaining the Method and Technique
of Administering Boiled China Root
Which the Invincible Charles V
Recently Employed

And Summarizing
Among Other Things
The Substance of an Epistle to Jacobus Sylvius
That is of Great Usefulness
To Students of the Truth,
Especially of the Human Fabric;
Since it Shows How Easily Galen
Has Heretofore Been
Excessively Trusted on That Subject.

There is Also Attached to This Epistle
An Index of Important Subjects and Words

Basel
From the Press of Joannes Oporinus
In the Year of Human Salvation
1546
In the Month of October

To the Illustrious and Great
Duke of Tuscany

COSIMO DE' MEDICI
Patron of Studies

Greetings from
FRANCISCUS VESALIUS[1]

Since Jacob Scepper,[2] a young man who in my judge-ment is outstandingly well versed in medicine and the disciplines that relate to it, came here for his studies and I often met with him concerning the affairs of our native country, I began to question him closely about what doctors were doing in Belgium, and whether anything had been published by them to assist and enhance our common studies, of which we had not yet been informed. He brought forth among other things an epistle of my brother Andreas, who is extremely devoted to your Illustrious

---

[1] Vesalius' younger brother Franciscus was the third child of Andries and Isabel Van Wesele. He studied Medicine at the University of Ferrara, from where this letter is dated.

[2] Probably a son of Charles V's diplomat Cornelius Duplicius De Schepper (1501–1555).

Majesty and a great admirer of your virtues for which you are so praised to all the world, and he is a keen herald thereof among the learned and all his friends. Scepper confirmed that he had written out this epistle from the original, and entrusted it to me with many names so that he could say it was circulating among the Belgians variously written out in the hands of certain people and be thought no less worthy to them than to himself, so that it would be set in type and become common to all. Since it appeared likewise to many people ✳ to whom it was shown, I did not hesitate to ask Scepperus to make me a copy of my brother's work to send immediately to Ioannes Oporinus,[3] the painstaking and highly educated printer, formerly a professor of Greek literature, lest it be badly printed through the negligence and greed of some inferior printer. I know well how great the good feeling of Oporinus is towards my brother and with what workmanship my brother's writings come from his press. The books of *De humani corporis fabrica* and the *Epitome* based upon them are no small credit to him and to our family, the Vesalii, and I wish the *Epitome* had not been spoiled so disgracefully by a certain Englishman (who I think lived with my brother for a time).[4] He took what had been written with great care succinctly as a list in the *Epitome* and expanded it with excerpts taken from the books of the *Fabrica* of which it is a summary. He utterly corrupted what had made it most praiseworthy and so roughly and absurdly copied what had been set forth with elegant drawing and engraving that he preserved no appearance of Oporinus'

---

[3]   Ioannes Oporinus (Johann Herbst), 1507–1578, son of the painter Hans Herbst, taught Latin at the Basel Latin School and Greek at the University of Basel before opening his own press. He had published Vesalius' *Fabrica* and its *Epitome* in 1543.

[4]   John Caius (1510–1573) records having lived with Vesalius for eight months in Padua. See O'Malley 1964, 101, 105–107. A Galen loyalist, he rejected Vesalius' skepticism of Galen's accuracy in human anatomy. Caius later enlarged the foundation of his former college at Cambridge, which was renamed Gonville and Caius College. Caius is wrongly charged here with plagiarizing Vesalius: the actual plagiarist was Thomas Geminus in *Compendiosa totius anatomiae delineatio* (London, 1545), see O'Malley 1964, 88 f., 223. See also n. 197 below.

majestic edition. It therefore seems to me no injustice if the author grieves that his name was not removed from that utterly incompetent English edition, to keep anyone from believing that such badly made and botched illustrations in the whole series of nerves and vessels had ever been produced by him. It is strange that that imitator, when plotting against the efforts of others with the itch for writing from which he suffered, did not read the epistle placed at the beginning of *De humani corporis fabrica* in which my brother wrote that he would willingly share the illustrations prepared at his expense with a diligent printer rather than have them badly printed and forced into a smaller format ✳ (which can never be large enough in any case).

As for the publication of the present epistle (which I should have thought should be titled a book had it not been written in the form of an epistle, however lengthy and varied, and which is seen to have grown unexpectedly beneath his hand into the size of a book), because it was not possible to make use of my brother's advice in that I feared it would be published by somebody in an inferior way before I could be sure of his opinion because of the intervals of distance and the crises that are now troubling Germany: I hope he will not disapprove of my effort and diligence, or even be angered because I arranged for publication (since someone would do that as well).

To avoid being deceived in my opinion, I judged that this labor should be performed for all students of medicine under the auspices of your immortal name and enhanced by its splendor. It is very clear to me that your Majesty holds my brother in the same esteem as do to a man all great men and lovers of letters. He hears easily and often from those with whom he now associates daily regarding what is widely circulated concerning the swiftness of your resolve, your unique knowledge of military affairs, your amazing swiftness (which was always certain) in those matters, and the sacred and never sufficiently praised government of your dominions, which should never be equal to your heroic spirit. So it is that among so many great and famous princes of Italy none is as often mentioned now by Germans and Belgians as Cosimo, Duke of Tuscany.[5] This is not only due to the civil and military gifts of your mind, but also because under your patronage letters and all the disciplines, which had been failing because of the neglect of many princes, not to mention the hostility of those who are counselors to them or given to them as advisors, are now seen to be nurtured ✳ and to grow in leaps and bounds. Indeed, we know how the family of doctors (to which you seemed the last remaining hope, had you not brought into the world wise, elegant and supremely esteemed sons, heirs of your virtues) has always concentrated all its energies upon the recruitment from everywhere of the most outstanding followers of the disciplines and their generous support. It did not concern itself in the past whether doctors were on the rise outside Italy and were held by other princes of Italy in the ranks of barbarians.

How much you strive to surpass your ancestors in this type of virtue is proof that the ancient university of Pisa, whose ancient splendor you wish with so much zeal and generosity to restore, lacks no effort on your part in supporting those who you do not doubt are leaders of their disciplines. Therefore it is no surprise that in so few years this university has begun such great advances to the great credit of all studies, and now shines brightest among the great universities of Italy.

Although critics are by no means absent who impress upon everybody the harshness of the weather at Pisa in that most elegant place in the world amid the greatest success in everything they do, nevertheless with your benevolence, though there was almost no need, an excellent provision was made for aqueducts which alone were found lacking, and you left nothing undone that would establish a pleasant home and market for the Muses there. Here too at the urgent advice of Francesco Campana,[6] a man distinguished for

---

[5] Cosimo I de' Medici (1519–1574) was Duke of Florence 1537–1574 but did not become Grand Duke of Tuscany until 1569.

[6] Private secretary to Duke Cosimo. See O'Malley 1964, 451 n. 54.

many disciplines and virtues, a chief confidant of your Majesty and no less zealous in your praise, Juan Bautista Recasulano[7] is the bishop of Cortona. For various reasons, and especially because of his incredible humanity and efficiency in conducting business, he is remembered with great fondness in the court of the Emperor, ✳ after he ceased to be the eloquent ambassador of your Majesty there. You will therefore be thought divinely given to the world for the strengthening and recovery of the disciplines. That is what everybody immediately began to foresee after your father Joannes,[8] easily the most highly praised commander in war of all in our memory (as he gave in no small measure to fate), when you were still very much a child took you from your nurse's arms and had you thrown headlong from a window higher than anyone could easily believe (were it not well known to everyone in Italy), as your genius was hurrying to the aid of your father. That was to determine whether his son, the one he hoped you would be, because you were taken up without any harm in his lap and in his mantle and were not to be torn apart into pieces, would fall to the ground. Warfare would tell the story, but only the greatest kind, as is foreseeable from your disposition.

Moreover, I believe my brother will be highly pleased to undertake this labor of his, a hazard of judgement, fortified by your authority. The proof of that is that I believe he will not reject the labor when your intellect is perfectly adapted to all tasks. For in addition to the use of new remedies, especially the method of administering the decoction of China root (which I see is given to those who are most devoted to your glory), and other medicines that are not unpleasant to know and are included in this epistle, reasons are added as well by which a devotee of truth can consider that Galen, easily the foremost of professors of

---

[7] In addition to his service as Bishop of Cortona, Recasulano was ambassador from the duke of Florence to Charles V, 1543–1545.

[8] Giovanni de' Medici or Giovanni dalle Bande Nere, 1498–1526, famed for his exploits as a condottiero or mercenary military captain.

anatomy, did not dissect humans but described other animals differing in many places from humans. Among other things, Galen did not provide true descriptions and uses of the parts; he often assigned functions incorrectly, and made many arguments in anatomy that are not altogether valid. One might well believe that all the things thus set forth by my brother ✳ are supplied as material to those who put their faith only in books, and think that Galen committed no anatomical errors whatever or any other mistakes, who have been stitching together writings for about three years now (unless they had started even before the publication of my brother's book). They refute in order everything he now investigates, one by one, as if they fell under common topics. They are occupied not only in praising Galen but in writing down calumnies or criticisms occasioned by a newly discovered truth.

By your generosity in the advance of learning, concerning which I have more than once heard my brother and many others taking pride, it has quickly been achieved that all these discoveries have been demonstrated to students at your university in Pisa by his dissections of bodies; and you yourself are aware how those who know the most in the presence of bodies, and the doctors and philosophers best prepared to correct Galen, have had to resort to my brother's opinion.

Indeed, some hope should dawn upon scholars that my brother will at some time return to your university and perform dissections of bodies for the profession of medicine. I should like to take refuge there as if to a free field which you provide for arguments about the disciplines – if only at last some will put themselves forward out of many (a number of whom have shamefully abandoned the enterprise), so that they may return without dissent from my brother's dissections just as I saw certain physicians come wholeheartedly into my brother's way of thinking when I once spent several months in Padua and was present when he was teaching anatomy to a very large class. Yet I too now, who was repeatedly put to the study of law by my parents contrary to my preference, which was inclined to medicine and always recoiled from the law, and by the many travels in which I gained nothing but a skill in various languages, ✳ being now for a long time dedicated I promise that I will pay attention to the diligent study of medicine, and in every way I will see to it that so far as I am able Vesalius will not be perceived as having been called away from his studies into the imperial household. I shall endeavor at some time to perform his role in dissections and for his sake I shall undertake to show to others the true facts he sets forth in the present work, and quickly and easily repel the sharp pens with which Sylvius (as if my brother needed to be deterred from his fine undertaking and hence from the truth) threatens him.

Since my brother now has little time at his disposal to refute the trifles of those who vainly strive to defend Galen in a matter that is self-evident and contrary to the senses, it is perhaps in no small way proper to my studies to fulfill my brother's duty and set forth much more copiously things that will stir the anger of those people in no half measure (since they disparage trust in reason and their eyes).

In the mean time I wish earnestly to ask your illustrious Majesty that he forgive this audacity of mine that I desire to act as advocate so freely and shamelessly on behalf of so many paradoxes and doctrines still quite alien to the ordinary study of medicine, and ascribe my brother and myself with all reverence to your Majesty in the number of your most dedicated servants.

At Ferrara, the third day of the ides of August in the year after Christ's birth 1546.[9]

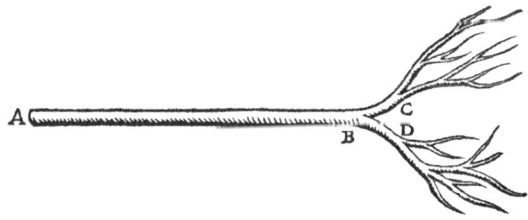

---

[9]   Misprinted M.D.LXVI, 1566.

To the Most Learned

JOACHIM ROELANTS,[10]

Chief Physician at Mechlin

GREETINGS TO A DEAR FRIEND

 have at last come to Ratisbon together with the Venetian ambassador Bernardo Navagero,[11] a special credit to his country not only for his famous knowledge of the disciplines but for many other reasons as well. Besides the presence there of close friends, the pile of letters accumulated in my absence diverted

---

[10] Born in Mechlin in 1496, Roelants entered school in Louvain and became a licentiate in medicine; he established a practice in Mechlin and served in the court of the Netherlands regent Margaret of Austria. He succeeded his father Cornelius as city physician of Mechlin in 1525, a post he held until his death in 1558. See O'Malley 1964, 454 n. 146.

[11] Bernardo Navagero (1507–1565) belonged to a patrician family in Venice and attended the University of Padua. Following his ambassadorship to the court of Charles V 1543–46, he served as ambassador to Suleiman the Magnificent 1550–52 and to pope Paul IV 1555–58. Named cardinal by pope Pius IV in 1561, he was bishop of Verona from 1562 until his death.

my mind from the weariness that I had to endure at the Emperor's behest in the difficult and grave medical crisis at Nymwegen.[12] It is amazing how much pleasure a quantity of letters from various places can bring; it is now a greater pleasure the larger the number there are to be read through together. I was delighted to read among the others two letters which I found written by you along with letters of your son,[13] a young man with the best prospects in my craft; you may readily imagine how delighted I was from the fact that they are the most trustworthy proof of your affection toward me, and they always have something to say about our activities and common enthusiasms. For this reason they deserve my greatest anticipation. Besides that, I was informed at length by your son's letters what the great defenders of ancient medicine (in my view) were up to in Paris and what was the success of their studies there.

While I postponed a reply to your letters because I was busy and needed to write my parents about private business, your letter was delivered to me asking me to write out the procedure by which a decoction of what is called the China root was administered to the Emperor and many in our court.

Occasion for writing about the China root

You also asked what my opinion is about that root, how successfully I use it, and with what recovery from adverse health. My friend Antonio Zuccha,[14] whom you praise in your letter for his

---

[12] Roman Noveomagus, mod. Nymwegen or Nijmegen. For the stay at Nymwegen to attend the illness of Navagero early in 1546, see Roth 1892, 207, and O'Malley 1964, 212.

[13] Martin Roelants, 1521–1598.

[14] Antonio Zuccha or Zucca is mentioned in the 1543 *Fabrica* for his interest in the larynx: "… a young man of outstanding talent, Antonio Succha, the rare and great hope of our common city Brussels, and indeed of all Belgium, on account of his singular knowledge of medicine and mathematics" (Bk. II ch. 22).

uncommon friendliness and well-known learning, is privately asking me the same questions. It is also remarkable how earnestly and carefully the German physicians of this region and functionaries in the households of their nobility are inquiring about the preparation of this decoction. They will not cease being a nuisance or put a stop to their prying and importunity until they learn the procedure by which they believe we administer that decoction. Besides certain things I will pass over in silence, they were so keen to gain this information that they have seen fit to extol the potency of this decoction to their princes so that the princes themselves have not hesitated to press home their entreaty on their doctors' behalf in the Emperor's presence in the hope that the entire method of administering the decoction would be explained by us to them.

With what success many have used the China

So great was the fame and reputation conferred upon the medication in a short time by the prestige of the Emperor, who took the China decoction more on his own initiative than on the advice of Dr. Cornelius[15] (whose services he principally employs since the departure of Dr. Cavalius).[16] While I was still visiting the sick under Venetian instructors and the principal doctors practicing there, this root was favored there with the greatest expectations and highest praises. But it was put to use in one case after another with little success. This was the result not so much of a fault in the medication or inexperience in its administration as the fact that it then began to be

---

[15] Cornelis van Baersdorp (ca. 1485–1565) of Bruges, second-ranking of the Emperor's physicians after Narciso Vertunno. See O'Malley 1958, 473 ff., and 1964, 194 f.

[16] On Dr. Caballus, "a physician of no particular merit [who] possessed the knack of gaining and holding the emperor's confidence," see O'Malley 1958, 478 ff., and van Male 1843, 14–15, 22–23. Caballus did not, in fact, leave the service of the Emperor. He is reported as still in Charles' medical retinue in 1551.

given to patients who were otherwise near death and in whom no cure could be expected by a doctor. ✳

13   The first patient to whom I saw a decoction of the China administered, had it sent from Antwerp together with an empiric who said he knew its use well and had administered it successfully in Portugal. The patient was suffering no symptom of the Gallic disease[17] that was not severe, and was extremely wasted as if bearing death in his jaws. As soon as a sweat appeared to be brought on by virtue of the China, urinations were moved at the same time and he took less nourishment than usual; he immediately expired. When his internal organs were examined at the request of his parents and friends, they were in such a state that no reason could be imagined to explain how life was prolonged for such a time in a person. Another patient who was being given the China decoction about the same time was seen to have the Gallic disease without any damage to the skin but with extreme thinning or emaciation of the body; he had developed significant stones in his left kidney. After he had used the decoction for about ten days the doctors recommended that it be diluted because of the extraordinary power they believed it had to liquefy and produce urination and sweating; after a few days he died with extreme pain in the kidneys.

After use of the China had been ridiculed and made light of by everyone more on account of serious harm to the sick than because of its virtues, and was virtually unmentioned, another physician came from Antwerp solemnly declaring that the China was a unique and sacred medicine and saying that there was no illness however deplorable

---

[17]   Syphilis, called the "Gallic Disease" because Charles VIII's French mercenaries spread the infection following their two-and-one-half-month occupation of Naples in 1495. The French called it "the Neapolitan sickness"; the Italians called it "the French sickness." See Claude Quétel, *The History of Syphilis*, pp. 11 ff. The modern name derives from Girolamo Fracastoro's poem *Syphilis* (1531), which popularized the doctrine of "seeds" of disease. See J. Arrizabalaga, J. Henderson, and R. French, *The Great Pox. The French Disease in Renaissance Europe* (Yale, 1997), pp. 244 ff.

for which it could not bring relief. The nobility trusted him so much (since he could also be credited with some erudition) that the China was given to the bishop of Venice, who lay ill with a wasting disease and unknown maladies. But before he had fairly begun to use it, the man, worthy of every praise, passed to his ancestors. So it was that so long as I was in Italy the China decoction had been fully rejected; nor did I, following my teachers, make much of it. *

I hear that at about the same time a fellow student in Burgundy   14 had employed a China decoction prepared with wine and great promises. But such was its success that since then the name of the China has scarcely been heard. A year ago, however, the most famous man in the world, Jean-Baptiste Gastaldo,[18] was bedridden much of the winter in agony of a nervous type below the loins and suffered a kind of paralysis together with a weakness of the stomach and appeared restored to health before the beginning of spring. After he had begun to be up and about, at the persuasion of some friends he used China decoction with good success. There was likewise the Spanish nobleman who you knew was brought from Mechlin[19] to Brussels: he made the China more praised than many other treatments. Also at that time four or five patients who had the Gallic disease urged the doctors to treat them also with the China decoction; in some of them we saw a very praiseworthy outcome. In others, though, who were more gravely

---

[18] Count Palatine Giovanni Battista Gastaldo was one of the Emperor's two Quartermasters General in the Imperial army, with Stefano Colonna in the Smalcalde War of 1540. "Gastaldo is esteemed a very practical man, and one possessed of correct judgement. His naturally good understanding has been improved by a great deal of experience … He discourses well and eloquently about the things he has seen; which are many, being a man of fifty years old. Yet his speech sometimes reveals too freely what is on his mind, seeing that his long services to the Emperor have been but inadequately requited." – Address to the Doge and Senate of Venice by Bernardo Navagero on his return from Germany in July 1546. Bradford 1850, 451 f.

[19] Mechlin was the city in which Roelants served as City Physician. The unnamed Spanish nobleman may have been Ludovicus Sanches (see n. 25 below), though his treatment was apparently not entirely successful.

affected with the disease, we observed a much lesser outcome than we should have hoped from a decoction of guaiac wood.[20]

You will likewise be aware that after setting aside the China the Emperor took a decoction of guaiac for his gout[21] and poor physical condition. There is no end of praise for novel remedies in certain quarters; several nobles of Spain were declaring to the Emperor and many other leading men that nothing is more perfect to prescribe for all diseases than this China and emphasized that guaiac wood is totally rejected in Spain and some parts of Italy, citing letters from their friends as well as various others who swore to the same view. A certain passion to take the China decoction came over the Emperor; he was more easily drawn to it because the China was so much recommended: the duration of treatment was brief, the dietary regimen was much easier than with the guaiac decoction, and it was prescribed with less commitment to rules (so to speak). So it was accomplished that people who are not present, knowing that the greatest ruler in the world used the China, have magnificent feelings and judgments about this remedy; they think they will be missing many things if they too do not also learn the system of administering this decoction. However, an argument for recommending the China would be of little effect coming from the Emperor, since he had used the China for only fifteen days, with an inconsistent diet, and with a frequently altered manner of dosage. Though he did not have difficulty breathing, nor for periods of time agonies of gout, the joint by which the left humerus is attached to the scapula for the past year felt somewhat impeded in the motion by which the arm is raised by the power of the deltoid muscle but now feels quite free. The same

---

[20] An oil or gum distilled from guaiac wood (*Guaiacum officinale* L.), introduced to Spain from the West Indies in 1508 and made popular in 1519 by Ulrich von Hutten, guaiac was used as a treatment for syphilis.

[21] *Articularis morbus* could be any form of arthritis but will here be translated as gout. On Charles V's attacks of gout, see Tyler 1956, 71; O'Malley 1958, 471 ff.; van Male 1843 *passim*.

problem was resolved in the left leg: during the same period, next to the ankle bones and the joint of the talus with the tibia, motion was quite impeded and seemed in general to mar his elegant form. In addition, when the Emperor was in the mood he had when he felt well, he was seen to discontinue his decoction under the pressure of urgent business, and in autumn he more or less decided to resume it with greater exactitude. At present he possesses the health that he especially enjoys when in the army and that all good people should wish with suppliant prayers that he have continually for the sake of the whole world. You know how well he thrives then, in the exercise of his full skills and supreme occupations. Consider also that after he takes the China he usually arises earlier than usual, sometimes for hunting before breakfast or riding by horseback to the most pleasant places that lie hereabouts, and taking lunch at not such a bad hour, for example after noon (as he often does), so that now a longer interval passes between his bath and a heavy lunch, which is a miracle to many, and the very light supper he is accustomed to take. But precisely these reasons reveal at least a good concoction if nothing else, ✳ that tends for various reasons to be spoiled from time to time. But no doubt you have heard more than once from the Imperial physicians, with whom you were previously associated, what the Emperor's style of life is and how much attention he pays to medical instructions. I am therefore not surprised that Dr. Cornelius was severely vexed with those who were responsible for the ways in which his diet[22] was to be established as a precaution against the volume of excremental fluids; by this diligence provision was to be made for his inner organs so they would perform their task faultlessly. Finally, if any excess humor should be accumulated, how it should be expelled by the attending portion of their art before descending into his chest or joints. They also looked after ways in which strength should be

---

[22]   On Charles V's unhealthy diet, see Brandi 1939, 560, 622; Tyler 1956, 271; O'Malley 1958, *passim*.

won over for the joints lest they take on an onset or influx of humors or transform otherwise benign nourishment that has been delivered to them into malign juices which the joints afterwards do not expel. They also mutter quietly among certain great men about how by diverting a humor ulcers can be created in some places by means of caustic medications, and they promise perpetual immunity from difficulty in breathing and the most certain illness of the joints – but they also publish these matters in pamphlets written to the Emperor at the pretext of some prince, just as if such matters were unknown to those whom the Emperor has put in charge of his health, and he were subjecting himself to medicine to be so unhappily troubled. It is a good thing that as long as he was still present Dr. Cavalius would urge the Emperor to be wary of bringing in such smatterers who would create a danger as great as precepts of such medicine might be useful. He would explain how much easier it is to make elegant speeches or write books about such things in the schools than to put them into actual daily practice on princes (who also claim power in medicine for themselves); this is especially true for gout, which sometimes severely infests all joints in the body. However, this topic leads in a direction other than the China decoction. You can easily guess how much ✳ praise for this decoction can be derived from the health of the Emperor. The celebrated Lord de Bossu[23] may well be consulted about this: it is seen that in taking this decoction he closely imitated a certain prescription[24] sent to the Emperor (from which a sort of formula for administration to my patients has been taken). It should be gently employed only for the drying of the body, and for strength, namely for gout and weakness of a neurological kind,

17

---

[23] Jean de Hénin-Liétard, (aka le Grand, 1499–1562): childhood friend and Grand Equerry of Emperor Charles V, who made him Count de Boussu in 1555. Scion of an aristocratic family in Hainaut, Belgium; his castle was visited by Charles V in 1545 and 1554.

[24] A marginal note adds "We have placed this text at the end of the Epistle."

though the patient may have suffered pain in the joints for months before use of the decoction or immediately before. The tremor of the hands seems to have been familiar to the Emperor for many years, about the same after using the decoction as before. When Lord Ludwig Sanches, the ruler of Sicily,[25] a man outstanding not only for his many rare natural gifts but also for a rare knowledge of disciplines and facts that all may observe, clearly experienced weakness of the stomach resulting from cold and moisture, a constriction of the veins weaving through the liver, obstruction of the bile ducts, and a phlegmy deflux from the head (as he has a quite uneven bodily constitution), he read through praises of the China decoction and being that way encouraged by friends, wished to have it prescribed to him. But in his case the method of coction had been quite altered (because water seemed to harm his stomach) and a drying diet was begun combined with the use of opportune purgations, though I should have wished a man worthy of the most perfect health to attain a more desired end. Nevertheless one of his close friends, to whom he wished the decoction to be demonstrated at the time, appeared to be quite well cured of the Gallic disease. But before administration of the decoction I had first employed sometimes venesection, sometimes a boiled-down poultice of Mesue[26] altered as usual with a quantity of hellebore[27] and a measure of milk whey, as I had also used other appropriate purgations in the employment of a decoction and a light and quite drying diet as we usually approve in the taking of a decoction of guaiac.

I achieved the same success in a number of other cases; but they wished to use the China decoction more on the advice of their friends than on my recommendation, and they did not suffer from

---

[25] Ludovicus Sanches (d. 1549), a Spanish nobleman, was protonotary or chief legal officer of the king of Sicily.

[26] Ioannes Mesue the younger (d. 1015), author of an *Opera medicinalia*.

[27] Any of fifty species of the genus Helleborus, most likely the species collectively known as "black hellebore" used in treating gout and other diseases.

bone growths, tumors, or malign ulcers, for which I am very certain the China decoction is far worse than a decoction of guaiac wood. Moreover, everything turns out better with the China decoction the more closely we imitate the procedure that we employ in the presentation of the guaiac wood, in a manner largely unchanged from the way the China decoction first began to be administered, unless perhaps someone should appear who is filled with bitter bile, in whom the constitution of the body and the disease are systematically considered. I have said these things so you will understand the extent to which I can respond to your request, in which you wish to have the method prescribed by which we administer the China decoction. I am not ashamed to admit my ignorance to you, particularly with this medicine, because not so long ago I dared to dispense it without a system, given as if by hand but led by bare trial and error. Until now I could not even be certain of the root's name, as the root itself is known in other ways.

Description of the China root

By some it is called Chyna, by others Chynna and by others Cyna, just as you only write Echina and it is called simply Achyna, as if an island or a place in India or in the recently discovered world[28] had supplied its name. It is imported by the people who bring in pepper, cloves, ginger, and our cinnamon: Portuguese and people sailing under the patronage of our Emperor. Those people say it is gathered around the shores of the sea, and as is likely in marshy areas of the sea, just as we observe roots of various species of reed and similar stalks growing. In fact if we rightly consider such roots, after they have been pulled from the earth for some reason or another by sailors or fishermen, or for any other reason are split into various fragments of uneven size, then

---

[28] That is, the New World.

rolled in seawater for a long time and finally rest on sandy beaches, their appearance is very close to that of our China. ✳

I do not know what it could better be compared to than roots 19 of the stalks just mentioned which present themselves to view in this way by chance on the shores of the sea or some rivers. Those are perhaps darker, whereas the China is slightly reddish like common bitter root or what we call *galanga*.[29] Indeed, what is more like the China than this *acorus*[30] after it has rotted and taken on a dull flavor if you take away its size and hardness? The China is notable for its large, rough, and uneven pieces, and is a little more woody though also significantly spongy and like the previously mentioned roots when it grows it is definitely succulent like the kind shipped to us, and soon shows itself very dry and often infested with worms.

When this happens you may find merchants who to pretend that rot and worms are not present wrap the China in the ordinary Armenian bolus[31] of the pharmacies in the same way that we know ginger is covered in the glutinous earth that is reddish and offered for sale in shops, especially in Antwerp. I have questioned our mutual friend Gerardus[32] (as he is especially expert on stems as he is on everything else) whether in his legation to Turkey he had learned anything definite on the subject of the China. However, I was unable to learn anything else from him but the fact that the China is imported to Constantinople also and was employed by a certain Jew with less success than the hope and expectation of the sick. From merchants I was able to learn only that the China is found near seashores and that the

---

[29] Or galangal, a rhizome of plants belonging to the ginger family. Its medicinal uses originated from Indonesia. Hildegard of Bingen (1098–1179) used it in various medicinal formulas.

[30] *Acorus calamus* or sweet flag, a grasslike evergreen plant, was cultivated in Asia for its medicinal properties.

[31] Armenian bole (bolus armenus, bole armeniac) is a usually red earthy clay native to Armenia. It was used as an astringent against diarrhea, dysentery, and hemmorhage.

[32] Identified in Vesalius' Index with surname Vueldwick.

natives use it for scabies just as we use rumex or lapatium.[33] But there is no lack of people who to make the China more marketable solemnly assert that every kind of illness is kept from the natives by the China decoction. We single out the China for our purposes, just as we do rhubarb and other roots, because it is outstanding in weight and as far as possible in that very dry root, succulent and less beset with worms or rot, and more resistant to other damage ✳ from putrefaction which it was able to keep out of its spongy, damp substance while drying. We should be quite careful in selecting these pieces, particularly the smooth and thin ones, from the woody and coarse pieces and those that split open at intervals; though the first and second qualities, as we call them, are still unclear to me; these show how great a preeminence the medicine possesses, by which many recommend it to the public. For since we distinguish its strength chiefly from its taste, why will we say it has a tasteless and quite sluggish flavor? For whenever you chew the China, which is otherwise dry and woody in consistency, and grind it with your teeth, you will be able to confirm that it has no flavor; still less (so far as is in its nature but not that of drugs close to it) does it present to a person tasting or otherwise handling it any odor or anything oily, which others contend they have observed in it. It deserves promotion or praise in a much different way from guaiac wood or the medications that are indigenous and familiar to us; we would be able to commend those with many praises for the uses for which the China is considered a sacred refuge. No one is unaware how much hope would be placed in the present circumstance upon the lesser and greater roots of centaury, upon rhapontic, bell of helerius or enula, aristolochia, gentian, galingale, cinquefoil, sorrel, and upon roots of such plants together with caper root (if one wants to adopt something astringent like that), wormwood, hyssop, calamint, pennyroyal, chamaedrys, juniper wood, our spikenard, decoctions of

20

---

[33] Rumex or lapatium, the genus of docks and sorrels, was thought to have astringent and purgative qualities.

the inner bark of ashwood, and other simples of that kind, since we would not be as irrational as the sick themselves to marvel so much at those exotics and put them to use no matter how much they are at odds with reason and the method of our art. I am unable to comprehend ✳ how something that many have long since credited to the China comes to be ascribed to it, and with what new impulse it now begins to be credited with new powers as its fame grows. Presumably it is hot, possessing the faculty of opening; it notably brings on urination and sweating; it is found to be a consumer and dryer of redundant and harmful substances and any number of different fluids and is thus a cleanser of the blood; it possesses the virtue of assuaging and cleansing; it keeps the belly now loose and now, when it chiefly brings on urines and sweats, tight. It is a remedy for a stomach beset with phlegm; it drives off diseases of the liver and spleen in no small way; it is an effective protection for those whom the stone torments and breaks it up. It drives off gout, especially benefits elephantiacs, cures skin diseases, and is no ordinary help for fistulas and malign and otherwise incurable ulcers. It divinely heals the Gallic disease for both recently and earlier infected patients, heals the ulcers caused by it, and repairs their scars. It crushes agonies in any limb, dissolves tumors, and aids any that are about to suppurate by heating, then opening, cleaning, and producing a scar. It brings health to corruptions and abscesses in the bones, relaxes convulsed and contracted sinews, dries those that are slackened and softened, and heats sinews that have been numbed by the Gallic disease; it fattens sinews troubled by wasting therefrom. It induces a pleasing odor to bodies that are festering and corpse-like, removes bad breath, and protects those who have trouble breathing. It scatters lengthy anginas, drives off injuries to the brain resulting from the Gallic disease, and halts every kind of fluxion with an uncommon benefit. Finally, the China is esteemed for the same virtues as guaiac. In fact, it is preferred to guaiac for many more faculties, even those that contradict each other, since the previously mentioned dull taste occurs in the China without any indication of astringency. ✳

22   However much you boil it in plain water, you will not give the water a different flavor than you will notice comes from barley not cleaned of its husk and the smallest portion of a sweet root; the China decoction is slightly red, very much like a tawny wine or wine that has taken on a red color from long storage in a jar. So from the point of view of manifest qualities, from a decoction of this root there is less reason to hope for stimulation of sweat and urine and the virtues for which the China is today praised, than could be expected from barley water. In fact, to conjecture from this evidence, when the China is to be used on the sick a decoction thereof should preferably be offered to those whom we see involved with the slow fevers, acrid humors, and altogether bilious condition of the Gallic sickness. Only with these patients, so far as concerns the method of our craft, have I known any genuine success from the use of the China decoction. I am also convinced that the China was first praised for this use.

Although, my Joachim, the various and more or less diametrically scattered virtues of the China arise from this success, and while those who use its decoction (without neglecting the skillful use of medications that necessarily accompany its administration and an opportune dietary system) sometimes meet with the desired end, our gadfly which we too often avoid should not be absent: I am talking about the recondite and occult faculty which we say is specific and essential, by which we establish the fourth order of medicines. This refuge is so broad that it embraces everything; we observe nothing hidden and unknown that we do not refer to this category. What exists in the present that cannot be ascribed to this wide field of the China? And not only to the China but to any spongy and rotten wood? Consequently our reason should no longer consider only the powers of the China, since it easily allows

23   us to ✳ take refuge in the declaration that the China moves urines and sweats and removes faulty humors through sweat and hidden respiration, and we know that many purgatives drive out a select, particular humor. On these grounds there will be none of the powers that anyone can claim is falsely ascribed to the China by those who have instituted it.

Method of preparing the China decoction

At some point it will be time to relate the method of preparing the China decoction and the technique of providing it that was largely presented to me more or less by hand in an Italian text. I will accomplish this more quickly and with less trouble, the more it can be described by a man long learned in the practices of our art and the greatest help of his fellow citizens. As soon as we have employed a purgation of the bowels, emptying of the first veins and then, if the occasion demands, venesection and the appropriate digestion of redundant humors, and finally the actual purgation according to the accustomed method for the sickness at hand, we present a China (as I previously explained) that is more dense and less affected by worms or decay, and as far as I can judge fresher in appearance and less dry. A twelfth of this is split transversely with a sharp knife into the thinnest possible discs or pieces just as if one were to separate coins lying on top of each other in a cylindrical figure or divide up an orange or a radish on a table transversely into pieces. The pieces are thrown into a leaded clay vase or what we also call a vitreous jar that can contain more or less sixteen pounds of water for convenience in cooking; it should not have too large an opening, and it should have a lid that fits. Twelve pounds of water are then poured over the pieces; it should be spring water or be otherwise close to that, and be thought on the basis of good and special evidence to be excellent and highly praised. In observing a water source, aqueducts claim no small priority ✳ because water should not flow a great distance through lead pipes. For the China to be better steeped in water for twenty-four hours, the jar should be kept over hot but not burning embers for that period of time. The decoction should then be kept in a slow, continuous fire that is not smoky until a third has evaporated. This point is best reached at night or at least in the evening of the day before the decoction will be used. Soon after cooking the water should be strained through a linen cloth and poured into another jar; or it should be left unstrained in the jar

24

and only so much as the patient will consume at the time should be strained from the jar. However you do this, whenever the pieces of the China are free from the water in which they were cooked and slightly dried on the strainer or another cloth, they should be put away somewhere for the use you will hear they have in an apparatus of second decoction. For this purpose the jar into which you have drained the strained water or the pot in which it was cooked should be placed for a while on warm embers or wrapped in towels or cloths and removed a little from the fire to keep it from cooling for a while until there will be a use for the decoction throughout the day. It serves for only one day, and the decoction should be produced fresh daily as long as it is taken. Those who promote the China fear that if it is kept too long it becomes acid and is somehow spoiled more quickly than a decoction of barley, forgetting how much heat and how many aromatic powers they otherwise attribute to the China when they say so much about it.

Quantity of the first China decoction to be administered, and the time to give it

Eight ounces[34] of this decoction or a little more should be given in the morning so it can be delivered when hot, and the same amount four hours before supper, during which, as at the mid-day meal, the decoction should be served warm in place of a beverage. I take the time of dosage, as with the decoction of guaiac, from the natural intervals of the day, so that the right spaces will be established from the twenty-four hours. So that a hot decoction be served at the fourth hour in the morning, lunch should be at the eighth hour; it should be served again hot ✳ at the fourth hour in the afternoon, and dinner at the eighth hour. In other words, so that four hours

25

---

[34] Lat. *unciae*; an *uncia* is the twelfth part of a unit.

28

intervene between taking the hot decoction given in the morning and lunch; and between lunch and taking the hot decoction after noon, eight hours, and between this dosage and supper, four hours. Between supper and the drink presented in the morning, eight hours also pass. We might instead in certain regimens for a patient decide to extend the final interval and wish the lunch to be more coterminous with supper, giving the decoction an hour or two later in the morning.

How a sweat should be induced

When patients take the decoction in the morning, they should lie in bed covered just as if they wished to induce a sweat. After it comes on as planned they should be wiped off everywhere with warm cloths without uncovering the body, and when the linens are pulled off if necessary a clean and well-dried nightgown should be put on. To avoid having to pull the linens away from the sweat or move the patient to a cold part of the bed, we fold a large bed linen together four or five deep into the same narrow space so they lie beneath the patient's arms and legs lengthwise and wrap his chest. We then put another folded linen beneath the head and neck; its two ends cover the whole chest and a part of the abdomen, and the head is also pro-tected according to plan. Because these linens make no small contri-bution to raising a sweat, when wet they are so capable of removing a sweat that after a session of sweating the patient can rest for a time in dry linens that are still warm from being laid upon and then get out of bed and having tried the means that effect removal of any kind of wastes, can take his midday meal. When a decoction of the China is administered before supper, the patient should take to his bed, and the same procedures we said should be performed before lunch should be followed. *

You know without my saying so from administering decoction 26 of guaiac and from the initial abundance of serous humor that patients are eager for more lengthy and copious treatments during the first days than in those that follow. But decoction of the China is many times worse than guaiac in bringing out sweat and urine.

There have been some for whom decoction of the China removes no sweat at all, just as we try in vain to bring on a sweat with a decoction of barley when the humors are thick or the skin dense. Similarly it is known that even guaiac has sometimes failed to provoke a sweat for the same reason except after several days. In administration of the decoction before supper, I do not know why some doctors reduce the dosage when using the China or Guaiac, and why before supper it is administered differently than several hours after supper before the patient sleeps, as if they thought he would snore all night in a sweat. That is why the Emperor also recently exchanged this regimen, which was prescribed throughout the time when guaiac was used, for some method that he believed to be the custom of the empirics. For while in the beginning he also took his decoction of the China after supper, following the example of Lord de Bossu and certain others, he now passes over this method and drinks what he used to take before bedtime several hours before supper with better results.

Those who prescribe use of the China have declared that a measure of a cyathus[35] of the decoction be administered in the morning and after noon; in place of this the Emperor took more or less ten ounces. But since this decoction of the first and second quality is scarcely better than a decoction of barley, I have guessed that it harms the stomachs of some patients, and have taken care to boil in an ounce of the China with six pounds of water until two pounds evaporate so that when this is done five or six ounces of the decoction sometimes are enough. Sometimes I have prescribed that

---

[35] The cyathus, originally a Greek wine-ladle and an Attic measure of about 1/12 pint, was later a type of wine-cup which was a standard unit of medicinal measurement in the 16th century.

two or three ounces of the china be boiled with twelve pounds of water to see whether a decoction from this process acquires more strength and power; for I think that such a small quantity of weak China root is boiled in such a large amount of water because it is sold by weight to many crowned heads. Although others have made this comparison when I recommended they make the same test with their patients, they disapproved this system of variation because I was trying to force a medication accomplishing so much with a little known power to a different end than others testified their experience recommended. But the pattern of variation has not so far seemed a hindrance to the method of guaiac treatment, as I am compelled to say it is with the China. For the same reason the air about which users of the China debate should be mild. The reasoning is taken from those to whom guaiac applies; those who first began to prescribe the China do not recommend the complete interval in a room that is closed and lacking in free air to breathe or without ventilation. In fact they allow patients to go out after the seventh or eighth day provided they walk moderately dressed and out of the wind and it does not disagree with their bowels. But I have so far given the China to no one who has not stayed in the same room or at least not avoided fresh air. This is not to say that so much power lies in thinning the China that I think because of the open pores of the skin air seriously disadvantages those using the China. Rather, I believe the tepid air and plenty of clothing keep the body readier to sweat and aid the weak power of the China. It is also for this reason that not so much calculation needs to be made of the time of year when it is better to take a decoction of the China as is needed in the use of guaiac; that is because we need not so much fear a disturbance of the air after drinking the China decoction, when the skin is somewhat rarefied.

Moist food is recommended first, being suited to acidity of the humors and great dryness; this includes young poultry or capons boiled and fish if any good ones are available and the patient has a

28 strong appetite for them. They have determined that bread not be reheated or, as we say, * twice cooked, but otherwise food should be cooked in the common way but without the seasoning of salt, just as they want salt left out in boiling meat and in every procedure. They prohibit ingredients containing vinegar and acids, the frequent use of sauces, and herb flavorings. They have also made their prohibition absolute for the first fourteen days and rule out for the remaining time all meats, which we approve in a drying diet. They permit seasoned items that we have in our pharmacies as preservatives, but only ones that sweeten; to these they add the flesh or juice of quinces prepared without the admixture of any aromatic. In a word, everything they recommend is altogether alien to a drying diet except for honey, which is the only such thing they approve of eating. I think they recommend honey prepared in the way it is said to be cooked by the Spanish, familiar to the Emperor at supper in winter, and which you know is quite beneficial against difficult breathing. The weight of bread, meats, or other food is not precisely determined except that they prefer a light diet, and the more meticulously it is established the happier the outcome they say can be expected from the China, adding with the greatest truth that no less a reward is conferred by a diet of this kind than is gained by the China decoction itself. For my part, with respect to the sourness of juices I cannot recommend this dietary method with any but the greatest approval. But because I am more often compelled by my patients themselves to administer a China decoction and a drying and extremely light diet is altogether necessary for them, you will quickly realize that I had to alter the regimen; from the entire prescription I therefore had recourse to sweat baths, the slightly acid raisins, almonds, pine nuts, and toast, but not keeping honey that has been cooked, pulled, and drawn like sugar sticks.

29 I shall not write down what weight or quantity I used of each ingredient, since you know * that in this respect much has to be assigned to the nature of the disease and the habits and strength of

the patient. In addition, when the decoction is first taken and just before the end of treatment, much differently from the middle of treatment, nearly everything should be altered, except that on the days when purgative drugs are administered patients should be indulged to a great extent; and you agree with me that the doctor's judgement should be free in no matter as much as in the weight of food.

What drink is useful

The drink, as was mentioned in the preparation of the decoction that we mentioned, prefaces the decoction itself, which is presented in sufficient quantity lukewarm at luncheon and supper and whenever an otherwise troublesome thirst is present. Though the originators of the China wished patients to sleep as if their strength or other occasion did not produce much effort, wine diluted with a decoction of China should be served at lunch and supper. In accordance with this rule the Emperor would generally take a first drink or swallow of wine mixed in this way and the second of the decoction by itself. Because of weakness of the stomach on account of cold, Ludwig Sanches,[36] whom I previously mentioned, appeared to be heavily affected by so great a force of water that wine also had to be given to him, sometimes after he had first chewed a portion of a medication consisting chiefly of three kinds of pepper; sometimes the medication came from the juice of quinces in a formula described by Rhazes where the medication was at first omitted from the book *De sanitate tuenda* and afterward at the urging of his friends added to the end of the book of Galen.[37]

---

[36] See n. 25 above.

[37] Like some other sentences in the *Epistle*, this one is murky in the Latin. *De sanitate tuenda* in six books is a work of Galen. The Persian author Rhazes or Razi (865–925) is described here as if he had edited, translated, or annotated the much earlier work of Galen.

Sleep and wakefulness

Those who take the decoction for sleep make the best use of night-time if they begin medication about two hours after supper. Those accustomed to sleep in the afternoon and who feel no harm therefrom, do not have sleep entirely denied them also at that time; so after sweat has been wiped away and energies more or less relaxed, a great many get a quiet sleep. ✳ You will thus easily guess that different kinds of sleep are recommended not so much for the sake of habit as because of the diversity of illnesses to which patients are subject.

Movement and rest

You will learn from those to whom we prescribe guaiac that exercise and rest should also be varied according to the temperament and makeup of those taking the decoction.

Concern about bodily wastes

Again from the use of decoction of guaiac we are also concerned about the purgation of wastes. We have already discussed the stimulation of sweat, which generally comes only or in greater quantity on the first days, and with which the impurities, which we know adhere to the skin after their unperceived expulsion, easily soften and are washed away as if in a bath. So far as concerns the purging of impurities of the eyes, ears, nose, mouth, and teeth, massage of the hair, which is the action of soothing with a comb, and washing of the face and hands, nothing is changed from the custom which one observes in time of health from the teaching of preceptors who preserve health. Excretion of urine is not impeded by use of the China, though it has less of a faculty for removing urine and setting it in motion than I hear attributed to it. At the start, as also in the use of

30

a decoction of guaiac, when as I was saying the sweats come out in the largest quantity, the urine is redder but presents its native color with the passage of time. The greatest care must be taken that the stomach respond as it should. In some patients it tends to be more compact, especially at the beginning and soon after purgation and when we perceive the sweating and urine are being drawn out in greater quantity.

At this point I cannot sufficiently admire those who were the first to prescribe a decoction of the China, that they wanted to assist retention by the intestine and directed that individual decoctions of the China, which should be employed at the time when the stomach is contracted, be cooked with half a dram of celery root together with the China; that is how I understand "half of an eighth of an ounce," which we read in their prescription. As celery root brings some assistance in stimulating urine, ✳ it also dries the feces of the 31 intestine more rapidly and makes the intestine less prone to excretion. It is equally admirable that they recommend that injections or enemas be introduced into the intestine; these are not brought from the shops of pharmacists, as they write, but from an extracted liquor or water of chicory or borage, with a small quantity of rose oil added, or if that is not available, prepared with ordinary oil together with salt. It is not at all different than if they had suggested that every drying and sharp medicine should be avoided like an enemy and their special goal had had a view to the acidity of humors. So far I have not had these waters injected in a patient; but to conjecture from the guaiac decoction, I would order a decoction of the China to be introduced into the intestines with a coarse sugar, sometimes together with honey made from roses, and common olive oil (unless it happened that a calculation called for something medicinal), whenever I believed it opportune that the intestine should be evacuated with a clyster. Because I have found the power of the China to be so senseless in the first faculties, I have more than once provided it reduced to a powder to be introduced into the bowel in the same manner that we mix a powder of guaiac for the same purpose to some other patients who are suited to this treatment, and provide it as a purgative drug.

What affects of the mind are applicable

So far as concerns mental affects, I want all who use the China to have an active life and to keep away all grief and every care in that solitude and circumspection comparable to a prison, though I am not unaware what a difference in the treatment of diseases is made by torpid, idle, relaxed emotions as distinct from those that are large, agitated, and heavily occupied. The presence of friends greatly helps those who are using a decoction of guaiac, especially those suffering from the Gallic disease; long conversation about pleasant topics and subjects that entertain the mind; likewise sport, with which one can restore oneself without great mental concentration ✳ or a discerning anticipation of successful outcomes. You will find certain games like the robber game[38] which I know reduce the strength of the primary faculty no less than the most arduous and difficult pursuits in the disciplines.

Sexual activity

When I first chose the method of administering the China decoction, I could not be sufficiently surprised that no one employed sex; women who could have aroused to venery were kept at a great distance; this was often repeated. Reasons occurred to me, however, that would persuade those who employ guaiac to think about other things besides sex. The sharpness of the humors that accompany the Gallic disease and the temperature which is known to stimulate

---

[38] *Latrunculi*, a Roman game of "robbers" somewhat resembling checkers or chess (OLD).

bilious and melancholic men to excrete semen seem a less effective argument than one could make. But when I put the China to use for a certain length of time I believed that precept should not be casually ignored, for it is astonishing how much men who drink the China decoction have erections. This is especially the case if a diet that is not thin, drying, and especially choice is begun at the same time, but patients are fed as if in a rapid relaxation of the mind, without drying, with much reclining on the back. It is likely that in this case a decoction of the China, which is otherwise so spongy, can accomplish something by its dampness; not so for guaiacum and other substances that can be given with it for the same use according to method. I was therefore unwilling now to neglect this rule, especially when I knew that some, while employing the China decoction, were so aroused to venery that though they had long refrained from sex and had avoided it for various reasons, did not abstain from intercourse.

How long the first decoction should be used

I am well aware that you anticipate it will finally be asked how long we administer this decoction whose preparation I explained above. The rule from which I learned the use of the China sets forth an interval of twenty-four days. I have adopted this for my choice and judgement, just as I believe skill is located ✳ in the method and reasoning of each person managing a treatment. Since that time, I have not known others except the Emperor who have used this decoction. In fact, I have thought the treatment should be extended by more patients for more days, hoping for an accordingly better success, since I could not at other times oppose its use. Just as I previously believed that before using the decoction the patient should be evacuated with appropriate purgations and if necessary with bloodletting, so too I observe that the purgation should be commenced for more or less ten days according to each patient's nature. Likewise we end a treatment with this decoction by a purgation; how seldom that is the same, you know well from your particular diligence in medical work.

A method of preparing and taking a second decoction

I believe you recall that I wrote when the decoction I just mentioned was to be strained, the pieces of the China which had just been boiled should be put away somewhere for another use, namely for the preparation of a new or second decoction. After someone has fully used the first decoction, after about ten days have followed, a second decoction, whose method of preparation I shall now add, should be taken at the time allotted for the drink. Two ounces of the previously boiled pieces that have been dried in the sun or shade are steeped for twenty-four hours in twelve pounds of a select water in the pot that we previously used and then cooked over a slow, smokeless fire until a little more than a third is evaporated. When the water has been passed through a strainer, it is given as a drink, and the same decoction is duly supplied according to the prescription on each day. The duration of this use is not so long in the bedroom or even indoors, nor is the observance of a diet or stimulation of sweat or urine, as was the case with the earlier decoction. As those who prescribe the method of administering the China decoction say it is suited to remove ulcers extremely resistant to healing, they also have not neglected to recommend the method of use with which we employ guaiac to treat the Gallic disease when it attacks with many malignant ulcers: * they bathe the ulcers with a decoction of the China and cover them with towels soaked in the decoction. In this matter I have set out neither to praise nor disparage the China, since the occasion has so far not been given me to test it fully; I would therefore not provide it to any who have broken out in diseased ulcers or otherwise resist treatment and who are at the same time

34

extremely ravaged by the Gallic disease. I am not unaware that some have made little progress in a similar treatment for sores, and I have long known that treatment of a slight ulceration or a kind of sore behind the ear has been tried without success.

I also do not know what common ground there is between the juice of the China and the method of treating sores; I must acknowledge the China has no manifest qualities of cleansing or drying, though its qualities are also inconsistent. Still, it would be useful for me at one time or another to apply the China decoction to bodily members that are in pain one way or another because perhaps it has succeeded more as a warm fomentation than for another secret power. Just as I have not yet undertaken to cook the China in wine, so too I have not added another medicine to it, such as the root of sorrel, cyclamen, frankincense, lavender, and some others I have sometimes heard have been mixed with guaiac when it was hoped guaiac would accomplish some more extravagant special effect. For as guaiac has always worked well for me by itself, so too I have not believed it should be introduced in a new remedy.

A way of administering Sparta parilla[39]

To this category should definitely be assigned a remedy that is praised to the skies by many, particularly merchants, as have many who have returned here from Portugal. I know nothing at all about the type of stalk that I could compare to any plant of ours; the root is short and densely covered with nodes; it puts off from its lower side long twigs like continuous simple branches instead of roots. From its upper side grow a large number of stalks next to each other, which we could compare to dried-out shoots of hops, ✳ except that the cortex is more

35

---

[39] Better known as sarsaparilla (*Smilax regelii*), a popular treatment for syphilis when it was introduced from the New World.

or less divided into squares like the mulberry and sometimes appears without thorns. It is seen to grow in marshy places and to need a support to hold it up like viny plants. I have not yet seen its leaves or flowers: only the root with its stalk-like stems, of which more than twenty-five grow from one root that does not exceed the length of a palm and the thickness of two thumbs; throughout its length it does not divide into branches. The whole stem was a forearm's length for ease of carrying or convenient gathering in bundles without breaking, and tied together like vine trimmings. If it is ever unfolded, it is about the length of a man if the tips are off. So far I have been able to see only the one stalk sent to our residence from Portugal; a label written in French was attached that indicated the stalk was called Sparta parilla (to the Spanish it sounds like a low mulberry[40]), imported from India with great usefulness to mortals; it cures every kind of disease, especially the Gallic, if one having first taken the right purgation in the judgement of the doctor in attendance and according to the nature of the disease suffered, cooks an ounce of this stem in two measures of water, and drinks a twelfth of a pint of this decoction hot in the morning and before supper, and finally takes it as the beverage at lunch and supper with no use forbidden of foods that we recommend to those who are otherwise healthy. In addition, to those who used the decoction, free egress from the house was granted and it was added that the decoction was a significant remedy when applied to ulcers and painful limbs with the help of towels that had been soaked in it. Near the end of the dosage, which was recommended to the twenty-fourth day, as well as at the middle of the time, a purgation was to be presented. Nothing further is written on the label, but it is perfectly clear that the writer or author wished to set down ✳ about the same method of treatment and usage that he knew is observed in presenting the China

36

---

[40] Sp. *zarzaparilla*, a small, brambled vine, a member of the genus *Smilax*.

decoction in the same place from whence we have this Sparta parilla sent over. You may say that these stalks or shoots are utterly tasteless, no less than the China; indeed they are far more lacking than the China in any manifest quality, which is justly wished for in medications of this kind. For that reason I have so far not considered their risk worth taking, since a person to whom these cuttings have been sent by friends instead of by divine aid will soon appear cured of the Gallic disease by a decoction of guaiac. I hope I may be permitted to leave these cuttings alone and take some portions of them to send to the friends so they may ponder these Indian impostures and offer them to other doctors.

Native and familiar drugs should be put to use rather than exotics

This is what has occurred to me to write about the China. We should take things of the kind that should rather impel us persistently to the native and familiar and whose effect upon conditions of that type is known to us, into our experience rather than advise patients to use those unknown, arid, insipid, and quite odorless roots or stalks which for a while come at a very high price. You would scarcely believe how successfully Stephan de Casala, the Emperor's surgeon, employed the root of quinquefoil, which we commonly call *tormentilla*,[41] on some paupers, giving its decoction for several days in the morning without any great dietary regimen or bed rest; ulcerations were promptly cleansed by this decoction and then covered with a common plaster which we call *triapharmacum* because it is made of vinegar, oil, and litharge.[42]

---

[41] Cinquefoil or tormentilla erecta, a common weed thought to have antispasmodic, astringent, anti-inflammatory, diuretic, tonic, antiphlogistic, antiseptic, and hemostatic properties.

[42] Litharge is an oxide of lead, PbO.

Decoction of Chamaedrys[43]

I also cannot condemn a medication that was given in its own time and praised by the nobility at Geneva like something divine; it was sent with the greatest promises of a permanent removal of joint disease a few months ago by Mr. Marzilius Colla, master of the Emperor's horses, who is worthy of better health because of the many mental gifts with which he is endowed. *

37   Now it has been administered to the Emperor in the same way, so that he is said with the greatest certitude (provided he shall have put this medicine to use) to be immune to joint disease. The entire description is, I swear, quite empirical and contains nothing other than a purgation to be carried out at the beginning; then it calls for leaves of chamaedrys or low oak that is still green to be cut up if available (though I consider it preferable when dried, as with other herbs possessing the faculty of drying), then cooked in white wine; one twelfth of a pint of this decoction should be given in the morning, three hours before lunch. It should be added that more is to be hoped from such a medication, the earlier it is taken before lunch. In the dietary plan, acids and salts are forbidden, and drinking of this wine is prescribed for sixty days. It is added that the distilled liquor of low oak, or water, is compared in strength with the wine (which seems ridiculous to me), and accordingly water should be given to those who dislike the decoction. The label sent here has nothing else in general except a remarkably pretentious title and a long list of people treated by it, who by using it have lived many years immune to joint disease and who began using it when it was delivered. Cardinal Doria[44] was a leading figure in this group.

---

[43] *Teucrium chamaedrys* or wall germander, native to Europe and the Near East, was used for the treatment of gout. The name *chamaedrys* derives from the Greek "ground oak."

[44] Girolamo Doria (1495–1558), son of Admiral Andrea Doria (1466–1560).

Dr. Ludovicus Panizza,[45] a famous doctor of my age, advised the Emperor at Mantua to use many medicaments, especially terebinth, of which a portion should be eaten daily in the evening for the space of a year, observing a suitable dietary regime at the same time and repeatedly taking purgative medicines. You are well aware how fruitlessly such precepts are written for the Emperor, as you have known his habits and way of living for many years now.

No small results can be expected from genuine rhapontic[46]

My colleague Dr. Gerard,[47] considering his particular devotion to simple medicines in otherwise severe ※ matters, showed me among other medicines that he took with him from his embassy to Turkey, a medicine called *rha* by Dioscorides[48] and *rhaponticum* by Latin authors. It is not the root which the learned count among simple medicines, greater centaury,[49] but a thinner root than the common rhapontic, redder, spongy, rather smooth, and not very odorous; such odor as it presents is pleasant, more or less like that of rhubarb. In a large amount of it there is scarcely a piece spoiled by worms or rot. When tasted it becomes sticky with a mild but quite noticeable astringency; when chewed it produces a pale or yellow color like excellent rhubarb; when broken it quite elegantly resembles the finest rhubarb with a complex

---

[45] Ludovico Panizza, a physician in Mantua, author of *De minuratione facienda* (1556, 1561) and other works.

[46] A form of rhubarb used medicinally, now chiefly for relief of menopausal symptoms.

[47] Gerardus Vueldwick (see n. 32 above).

[48] Dioscorides Pedanius, the 1st-century botanist who called this rhubarb ῥᾶ because it grew near the river Rha (mod. Volga).

[49] So called because its medicinal properties were said to have been discovered by Chiron the Centaur. 16th-century herbalists assigned the name *Centaurion maius* (by some confusion) to a composite plant or plants (OED).

and series of veinlets or red lines with white interstices, unless in this rhapontic a distinct and alternative color appears that is brighter and more pleasant to the viewer. Also, when taken it induces no nausea. So when those who have begun to administer guaiac and afterward have wanted the much-praised China are seen to suggest the strength of the medicines from Dioscorides when he lists the quite appreciable and much different faculties of rhapontic, I hope I would do what is worthwhile if in the treatment of Gallic disease I would sometimes begin to use this: especially since it is known to Dr. Gerard where, except from the mountains of Germany that lie close to us here, or from the common hypolapathus,[50] which (unless it is also rhapontic) is cultivated in the gardens of my country under the name rhubarb, one's own substantial supply of this rhapontic can be obtained. I for my part believe no slight hope can be placed in this rhapontic with the little risk that I have so far taken in treating the several conditions that Dioscorides enumerates. For this reason I think I shall write to you from time to time ✳ more about the actual method proceeding from our craft than I shall for the present about the foolish China or Sparta parilla. In the meantime, as earnest money there will be pieces of rhapontic and some stems of Sparta parilla which you will now receive with this letter since the China is for sale everywhere among the people of Antwerp.

But now the times are upon me when I must busy myself not so much with cures, where it is appropriate to bring rhapontic into our experience; instead I shall be compelled to make trial of terebinth, which has been introduced with the warmest recommendation by Gerard, and with gums and resins as well as certain metallics. For in the war which the most excellent and merciful Emperor has instituted to pacify affairs in Germany with great armaments but no less a mood (so may it be for all) that desires

---

[50] A cousin of rhubarb.

harmony and peace for all Germany, I am not greatly distracted by Gallic disease, obstructions of the organs, or chronic weakness (such as otherwise keep me worried most of the time and at length). I have instead dealt with bone fractures, dislocations, wounds, and other such matters which often attend such sport now, except for such anatomy as comes along by chance and is esteemed only by empirics. Seeing these things, I cannot resist now and again applying my surgical hands to a task; I am otherwise forced to keep my hands off in those conditions in whose treatment we perceive the true strength of medicine to lie. So distorted are the judgments of men who contend that the hands have nothing to do with medical work. Weight is given to this opinion by doctors who have nothing to do with medicine except with certain drugs that relieve the bowel, syrups that prepare humors for evacuation, and with several evacuating medications and some formulas for pricking and fomentations; these doctors increase the great hauteur with which they conduct silly arguments about which vein should be cut open in sickness. Moreover, these pestilential men ✳ are so dependent 40 upon calumny that if they notice that someone knows things they do not, they admit that he is indeed an expert about those things, but they deny he is a doctor (which they are themselves in name only). It is as if someone who is truly to be considered nothing less than a doctor will finally become a doctor; and as if a person more accomplished in the earnest study of one of the medical disciplines would be resistant to the remaining knowledge of medicine. I know to what degree some on account of expertise in languages, others through mathematical studies, others through the constant examination of simple medicines, have been exposed to scorn by the most ignorant of these, doctors gaping at nothing but profit who have put their studies far behind them, in the company of Princes to whom these experts have been commended on other occasions by those present.

Hapless people who gratify themselves by publishing something

You are also not unaware what a hindrance to me when I first came to court was the study of anatomy that had altogether lapsed in our time, and the efforts that I made affecting all medical students. But, my Joachim, the fortunes of mortals are as each one allows his fate to dictate,[51] and there is no one who does not think daily how he torments himself and wishes himself free of this life. What falls out of use with some, I believe is especially familiar to others who satisfy themselves by thinking about a subject and bringing it to light; and in this way those who do not know they should escape attention by being born and dying expose themselves to so many calumnies, and it is their doing that they daily hear about their writings things that hurt their feelings. I therefore bear it with greater equanimity that I live at court away from the sweet leisure of my studies and do not teach medicine at Pisa for a stipend of eight hundred crowns from the illustrious Duke of Tuscany Cosimo de' Medici, the principal patron of the now largely declining disciplines.[52] I am just now in such a position that even if I strongly wished to do so, and no matter how strongly a self-love urged me, I am unable to bring myself to consider a new project ✳ or think about a publication. To pass over others and talk about myself, so entangled is everybody only in authorities and so few students of truth does this generation have, and people follow the disciplines only through digests or more accurately time wasters, that I hear many are furious with me because in my writings I appear to them to have scorned the

41

---

51 *Viz.* Each man's fate is his own doing.
52 Duke Cosimo had invited Vesalius to Pisa to present anatomical presentations. See O'Malley 1964, 197. O'Malley places the offer in the late spring or early summer of 1542, and Vesalius gave his course with great success in January–February 1544 (Siraisi 1990, 164). The offer of a more permanent position with a salary of 800 crowns came after those lectures (O'Malley op. cit. 199–203).

authority of Galen, the prince of physicians and universal preceptor, and because I have everywhere rejected his opinions. Any error I have found in his writings, they say, I have reported. They are unfair to me and our studies, and likewise to our generation. Though they should have been glad for these things, because I was the first that had the courage to drive out false and untrustworthy opinion from the minds of men and to uncover the rare imposture of the Greeks, and because I provided a unique occasion for investigating the truth, nevertheless you will find many who take such a perfunctory look at my labors regarding the authority of Galen that they still argue (without examining the body) that nothing at all was written by Galen with anything but the highest truth. You have seen writings of many of the most learned men of our time who have publicly praised my youthful efforts far beyond their merit; they are happy to say that their trust is placed in their own eyes, not in the writings of Galen.

Occasion for the letter of Sylvius in which it was declared that nothing written by Galen is completely in error

But I understand there are many who even though they grant that I have provided something worthwhile and credit me with more than I recognize, nevertheless are grievously inflamed for Galen's sake. Among them I would have counted anyone rather than Jacob Sylvius, the finest of the doctors of our time, had he not in a letter sent via your son attested to that reaction with extreme ardor, saying that he had read my work *De humani corporis fabrica*. You may now readily understand what I included in my letter to him which I sent to you at Nymwegen to be delivered by your son to Sylvius. You wonder what pages I sent to Sylvius, and if by chance the letter ✳ was only to Sylvius, what argument it employed, believing 42 it contained something about our common interests, which you

47

strongly wish to have told to you as well, as you are altogether eager to know about my trifles. I responded to Sylvius in that letter and added at length certain things which seemed to deal especially with the agitated mood of his letter. But everything was for the most part anatomical, such as observations by which things that caused him displeasure were explained away and were meant at the same time to show why I was unable to descend to his opinions. When your son departed for Paris to resume his studies and I recommended him in letters to Vasses,[53] Fernell,[54] Oliveri,[55] and others whom I most respect for their erudition, I considered Sylvius' reputation among the best, as you also are aware. Besides my praise for your son I inserted other remarks about our common interests, adding as well that if any things in my anatomical book were unsatisfactory in his judgment, he should let me know about them; because I thought it relevant also to him who has achieved a great name in anatomy among his peers if something was published by me, inasmuch as I enrolled in medicine under him. So it was under this circumstance that Sylvius wrote a letter in which he indicated in a standard and more or less formulaic way that he did not claim so much erudition or authority that in so great a matter he either wished or was able to be the arbiter; however, since I had set him up as the judge he would briefly disclose his opinion. Then, writing unfairly that I was belittling Galen, he listed the items by which he was particularly offended, namely that I had announced that Galen had not dissected human bodies; that he had observed the veins of the hand only in living persons; that the work *De usu partium* was

---

[53] On Jean Vasses of Meaux, the dean and chief administrative officer of the faculty of medicine at Paris, see O'Malley 1964, 37f.

[54] On Jean Fernel (1497–1558), who first used the word "physiology" to describe the study of bodily function, see Sherrington 1946.

[55] Otherwise unknown. See O'Malley 1964, 47.

copied out of other books. Other things in addition were picked out of my preface where I assign censure to Galen because of doctors who follow him, none of whom has been found * who does not agree with Sylvius, since he writes that he still believes that nothing wrong has been handed down by Galen. For this reason he wrote in his letter that in all the places where I said that Galen had falsely or wrongly written his description of the human body, or passed over something, or offended in an untrue account of the use of the parts, or invalidly attributed groundless reasons to the strength of his opinions, Galen was totally free of any fault.

43

I should have expected such an opinion from anyone but him, because after being warned, and now again teaching medicine outside his usual way of thinking, from the beginning, when Anatomy quickly comes into the picture, he should have been so zealous for truth that although he had perhaps read the books of Galen out loud many times already without finding anything wrong, he might now have begun to scrutinize them differently, set aside his piety (as befits a lover of truth), and compared the things I was saying in dissections of bodies; or if because of his aged condition reading occupied him by itself more than enough, at least he should have asked his students to make an inspection. I cannot accept what he writes, that he took every precaution at the university to prevent the least suspicion from creeping up on the listener that Vesalius was singled out by Sylvius because he has great love for me, esteems me, and wishes to keep me as his friend provided I free Galen of false accusations and place the blame either on my youth or on the zeal of Italians who wish Galen ill; for he intimates that he and I cannot agree unless both of us agree with Galen, adding besides to increase his apprehension that even if he said nothing himself, the walls nevertheless would tell their opinion about my nighttime labors. Several of his students in Anatomy, he writes, who are extremely proficient in his point of view are sharpening their pens against me, being indignant that any fault has been found with the universal patron of physicians. He therefore wished to be told what I wish him to do in this matter on my behalf, how * I would like him to issue a retraction and avert the sharp pens of his disciples from me. Thus, my Joachim, you have the story in plain view; it did not call only for the dense and lengthy letter to Sylvius, but for a book: not only to refute his judgment but also to lay open to the most experienced students of anatomy (if it please the Gods) the argument by which they may at last sharpen their pens when they have undertaken a dissection with sharp scalpels as diligently as I have done and consider without prejudice things that were set forth with the greatest candor for the chief purpose of aiding their studies, comparing the places accurately; so that they should not be occupied simply in praising Galen and deceitfully avoiding matters that I wished to be common knowledge to students. But I am not so ignorant of things, nor did I so casually notice Titus Livy's praise of the Carthaginians or Homer's praise of the Trojans, that I did not think it should be stated in every way and as much as possible with my sterile and meager diction that Galen is to be declared so to speak the author of all good things, the rare miracle of Nature, and the prince of physicians after Hippocrates, the chief and leader of professors of anatomy. To such a degree do I hold this view, that greater glory has been set in store for me from those who come after, among whom I expect to find the diligence of Galen in so many countless places: Galen, the foremost of all who have come after him. Therefore nothing can or should be more pleasing to me than the praise of Galen, and I yield to no one in my piety to him and deferential regard.

Neither Sylvius nor any disciples of his whom he talks about, or whose names he has decided to abuse, should have occupied themselves in reeling off Galen for my sake. Things falsely attributed to him that they esteem because of the tender devotion we excessively preserve toward authors, should rather be recalled, put onto the anvil, and judged with exact and tireless diligence so that by this means others may be spurred to the study of truth and I may be clearly warned if I have not made sufficient observation. ✳

This should not be as if I had started a fight with Galen, and I should not grieve too much that we have been so tricked by him that I have been exposed to scorn and finally so overwhelmed by some common theories.

Occasion for the opinion, here to be recorded, of the letter in which Vesalius replied to Sylvius

I will gladly write down how I briefly refuted those theories in the letter to Sylvius, since although I have not saved a copy of my letter (I had replied quickly because of the opportunity for a messenger) I shall at least put its content as I do into the letters which I send to my friends from time to time concerning my activities or which I write to patients or other doctors about the treatment of certain complicated conditions that have some peculiar feature, in the form of a consilium,[56] as we call it, and sometimes I excerpt. First, I wrote that I was the less disturbed by the letter of Sylvius the more I had seen the most learned doctors and philosophers in the process of dissection say the same things as I have about Galen, and he took them with the same displeasure as he did me. They considered it incredible that this father of medicine had committed such errors in his Anatomical volumes, and that he had written them diligently and precisely, and with great self-approval, and it was believed even while he lived that what was in such an account more than in any other would always be true. But with the passage of time those critics began to grow softer, nor was there anyone in so great a number who went on to correct Galen when there was a body present and with the greatest reluctance

---

[56] Examples of such consilia and letters to patients are translated in O'Malley 1964, 378–404.

place greater trust in his own eyes than in the writings of Galen. So it was that I hoped Sylvius would change his opinion a little later as he gradually went on with his reading of my book, and for that reason would not exclude me from his patronage and love. But if it should undeservedly turn out otherwise, I wrote that whatever that was I would count myself among the calamities of the human race because I had heretofore not learned to lie and speak contrary to the opinion of my mind, especially when I perceive that the things which he blames in me are daily increased and in no way remove my age ✳ (which I know perfectly well has grown beyond the youth that was thrown in my teeth by Sylvius) from his opinion. I need not say why I should have been so disturbed that he offered some excuse in the pretext that the Italians wish Galen ill. The fact is that the divine genius of the Italians cherishes and reveres nothing so much as Galen, as they have abundantly shown by the publication of his works, though at the same time the Arabs do not entirely despise him, and they should not by any means be kept out of the hands of medical students. I should therefore deserve to be considered utterly impudent if I put the blame for my negligence (if any of the things of which I am accused occurs in them) upon such learned people who deserve the best from me and are so friendly to me. They no less than Sylvius now were accustomed to oppose my views when they first attended my dissections, nor is there any of them whom I had ever seen put their hand to a body; they themselves will know well that in this narrative none of them was my teacher or accomplice.

Galen did not dissect humans, but teaches the study of animals instead of man

I therefore attempted to show in my letter to Sylvius that although nothing more weighty or impressive can be said about so great an author, Galen did not teach the fabric of the human body, nor is it even likely that he had ever witnessed a human dissection. For though he writes that he had seen the dried-out cadavers of two men suitable for study of the bones,[57] he nowhere makes mention of dissected humans except for the body of a German soldier (whose dissection he did not attend).[58] It is known instead that his account is based on the bones of apes, not humans, as is obvious primarily in the differences that we see between man and the ape principally in the parts or places where Galen described the fabric of the bones in such a way that it has been possible for us to note certain differences.

A number of conjectures from the bones

I had next put together some examples of these differences by location in my letter to Sylvius as they came to mind, just as if I should now ✳ say on the basis of various passages that now come to mind that Galen describes a suture in his account of the bones of the upper maxilla that goes up from the middle of the perimeter of the eye toward the middle of the eyebrows and is perfectly visible in dogs and apes but not in humans; these differ in various ways from Galen's description of the eye sockets.[59]

47

---

[57] See *De anatomicis administrationibus* Ch. 2.221, tr. Singer 1956, 3. This is the Galenic text that Vesalius had edited for the 1542 Basel edition.

[58] See *De anat. adm.* Ch. 5.385, tr. Singer 1956, 77.

[59] This difference is discussed in Bk. I Ch. 9 of the *Fabrica*. "Galen had described the premaxillary bone and suture of the dog as though present in man and thus could not have been familiar with human anatomy. Vesalius thus opened a great controversy of singular importance in comparative morphology, which was to rage with bitter polemics for nearly four centuries and to be settled only in recent times. ... His discovery, from which he made the correct deduction, was one of the major factors leading to the overthrow of Galenical anatomy." (Saunders & O'Malley 1950, 58).

Moreover, another particular suture[60] is told of in simians and those animals that have long, protruding canine teeth, that passes between the canine tooth and the incisor nearest to it and is com-

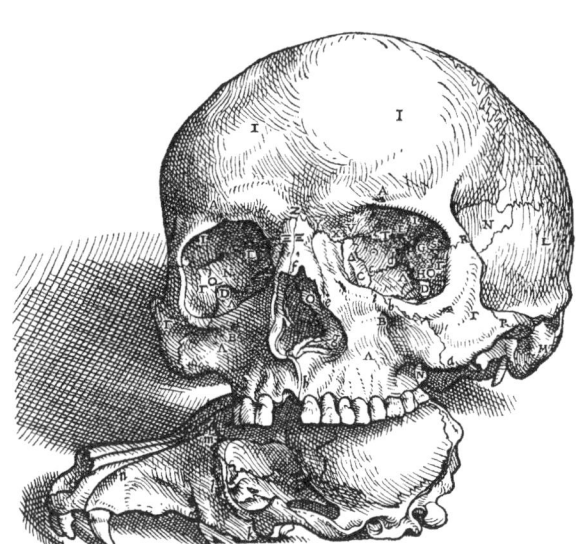

mon to a suture that runs transversely in the palate of those animals from one canine tooth to the other; since man, who has an otherwise very short jaw, lacks these sutures, we cannot with Galen ascribe to man a special bone in which the incisor teeth are fixed. In addition, Galen wrote incorrectly that the suture running to the canine tooth from between the brows is borne in a con-

tinuous path when in fact it takes its beginning from about the middle of the suture that ends at the outer side of the nasal bone.

In the foramina of the skull there is a major difference between man and the ape. But as Galen considers the differences too hastily, we can gather very few principles into our way of thinking from his account of the foramina. Yet at this point the difference in the foramina transmitting the principal branches of the carotid arteries is scarcely to be passed over, since it is seen that they are carved out transversely in man and by the great diligence of nature over a long course, whereas in apes they are more or less straight,[61] rather similar to the foramina

---

60 The suture marked n [*sutura maxillo-incisiva*] in the figure of the canine skull above and in Bk. I Ch. 9 of the *Fabrica*, where Vesalius disputes the claim of Galen that it is found in humans.

61 In his legend for the fourth figure in Bk. I Ch. 12 of the *Fabrica* Vesalius refers to Galen's manual on the dissection of veins and arteries *De venarum arteriarumque*

that provide a path for the nerves there. Galen observed the fabric even of apes so casually that he represented it far differently than it is in fact or than the artifice of Nature is seen to be in the course of the arteries, saying that the carotid artery has a common foramen with the third pair of nerves coming out of the brain.

The sharpness or edge of the heads of the occiput in the dog or simian bone * to which the first vertebrae are articulated is a proof that the skull of a human had not been examined by Galen: the human skull has those condyles protruding very little, extending from the back forward over a long course, meriting nothing less than the name κορώνη;[62] but the condyles of the animals just named extend noticeably downward, and for that reason are rightly given that name by Galen along with the other heads of bones entering sinuses.

The lower maxilla of a human can in no way be parted in its middle by any method of boiling at the top of the chin as Galen had seen a number of jaws of quadrupeds separate into two distinct bones, even sometimes before boiling. And if Galen tells us the maxilla of a newborn baby is formed from two bones joined by symphysis or a seam, let us say so for the sake of piety.[63] We shall soon convict him of negligence,

---

dissectione with the remark "I am especially surprised at Galen, who passed over this large foramen (among many others) as if it were not worth considering in passing."

[62] Gk. *korōnē*, anything hooked (the word is also used for the coronoid process of the ulna). This paragraph refers to a passage in Bk. I Ch. 15 of the *Fabrica* where Vesalius takes issue with the 12th book of Galen's *De usu partium* (4.1 ff), where in an unusually prolix and sententious introduction Galen states (wrongly, as Vesalius points out) about the joint of the first vertebra with the occipital condyles that "It is immediately clear that Nature has prepared these concavities and protuberances for lateral movements in both directions" (tr. May 1968, 556). Galen concludes his prefatory observations "No beneficent being bears malice over anything, but naturally aids and adorns all. So too, though I am not unaware that times without number this book will be treated spitefully and abused by foolish and ignorant men, like an orphan fallen into the hands of drunkards, I am nevertheless undertaking to write it for the sake of those few who are capable of reading and understanding it correctly and judging what is said." (tr. May 1968, 559–60). Vesalius' 1555 version would add to his criticism of Galen regarding the motions of the head over the cervical vertebrae.

[63] In Book I Ch. 9 of the *Fabrica* Vesalius disagrees with Galen that the human mandible consists of two halves joined by symphysis, like that of most animals, and can

because he did not say that the occiput of an infant, the bones that are knit at the sacrum, and a great many more besides are constructed of several bones in the same way, such as the vertebrae themselves.

From this I am more strongly convinced that Galen had not seen the bone resembling the letter υ [*os hyoideum*] in man, because this bone, besides the fact that it is very different from the bone in animals that is comparable to this character or rather the λ and is formed of a great many ossicles whose connections and shapes require no ordinary study because of several muscles and on account of the voice, you will find none of these (except in animals) that you can readily compare to υ, Y, or λ because it is not constructed of other ossicles.[64]

We observe that the transverse processes of the seventh cervical vertebra are always perforated in man, otherwise than in some dogs, nearly all of them. It is clear that Galen followed the opinion of other anatomists that it is sometimes perforated and sometimes not.[65]

Because Galen saw that in apes the tenth thoracic vertebra is taken up on both sides differently than in man and believed that the two vertebrae placed beneath it lack transverse processes, * he

49

---

be separated by boiling. See Galen *De ossibus* Ch. 6 (2.754.15–18, tr. Singer 1952, 771): "The bone of the lower jaw is not single, as one might think, for when boiled it too is separated at the point of the chin, so it is seen to be composite." Fusion of the two halves of the human mandible at the *symphysis mandibulae* is completed in the human by the second post-natal year.

[64] In Book I Ch. 13, "On the Bone Resembling the Greek Upsilon," Vesalius criticized Galen's nomenclature but illustrated two very different hyoid bones, neither of which is typically human. The left hyoid in his illustration has features of the canine hyoid apparatus, which is an assemblage of ossicles joined by synchondroses. Vesalius' distinction here between the animal hyoid constructed of multiple ossicles and the human hyoid which is a single bone was not made in the *Fabrica*.

[65] In *De ossibus* p. 758 Galen had written that the seventh cervical vertebra is rarely perforated. In the fifteenth chapter of the first Book of the *Fabrica* Vesalius had written "No vertebra of the human neck has so far come to my attention that did not possess a perforated transverse process. But I have frequently discovered processes of the seventh vertebra of dogs and apes lacking a foramen." (He would omit this statement from his 1555 edition, which adds that they are rarely perforated in dogs and monkeys and adds to his critique of Galen). But the omission of this specific

reported that it was likewise in man; but it is perfectly clear that the tenth rib is articulated to the tenth vertebra by two attachments, one of which belongs to the apex of the transverse process, which is why that vertebra does not lack transverse processes. The eleventh and twelfth vertebrae of the human thorax also have transverse processes, though extended much less than those of the vertebrae above.[66]

I will pass by the fact that Galen testified more than once (though falsely) that every one of the ribs is attached to vertebrae by two joints. But when we observe the vertebrae of humans, as well as apes, dogs, hares, and animals of that kind in the course of posterior processes or spines, as well as transverse processes, what could be judged more in conformity with the truth than that Galen had paid attention to the bones of those animals but not human bones? As he himself writes, we may see that the aforesaid processes of several of the lower thoracic vertebrae and all of the lumbar vertebrae run so obviously upward, otherwise than in man. Moreover, among the lumbar vertebrae Galen describes a particular process as a bulwark for the nerves that take their origin there from the dorsal medulla; it does not escape my notice in apes and dogs (whose lower thoracic vertebrae have that process), but I know very well that man does not have it.[67]

---

matter from the 1543 *Fabrica* suggests that Vesalius had subsequently developed his critique.

[66] In *Fabrica* I Ch. 16, Vesalius wrote "In the transverse processes, the thoracic vertebrae also vary from one to another, but not so much as Galen stated in the thirteenth book of *De usu partium* where he deprived the tenth vertebra and the two beneath it of transverse processes." See Galen *De usu partium* 4.78–83, tr. May 1968, 588.

[67] See the section of *Fabrica* I Ch. 17 titled "The extra process which Galen ascribes to the lumbar vertebrae," citing book 13 of *De usu partium* and chapter 10 of *De ossibus*. Vesalius comments "I have never observed this process in human vertebrae; consequently, in investigating it I decided I should employ the same system and method by which I regularly investigate features described by Galen that I do not find in human anatomy, and which I perceive to be other than he stated. My practice is to examine all of these in dumb animals, especially dogs (of which there has never been a shortage); these have quite often shown me what Galen described, or instructed me what he meant as if leading me by the hand. I was unable to find this

The difference between man and ape in the fabric of the sacrum and coccyx cannot be explained in a brief account. But because no small amount of guesswork is necessary on that account, it will be useful to explain in at least a few words that Galen neglected human bones. In man, a large bone that is quite complex in shape lies beneath the lowest lumbar vertebra. This is made up of six particular bones at most,[68] that are so attached to each other by symphysis or coalescence that in humans of more advanced age there is no appearance of a joint in the posterior surface of the bone. ✳

50   But in the anterior region they exhibit a clear attachment that presents something in common with the meeting of the vertebral bodies. That connection is more obvious where the bones are attached as if to bodies (to compare it to the vertebrae), for where they are built together on the sides as with transverse processes, they scarcely show the line where they are joined. This series of bones becomes narrower as if from a wide base downward, unless one were to argue that the transverse processes of the second bone extend more to the sides than the processes of the first. The transverse processes of the three upper bones have both depressions and protrusions by which a close fit with the iliac bones is achieved. The upper surface of the bone is articulated to the lowest lumbar vertebra in the same way that we know it is joined to the vertebra lying upon it, whereas its lower surface rests upon a tubercle that is proportioned as if to the body of a cervical vertebra. Without another process it enters the depression of a succeeding bone, which with the three ossicles attached to it in order we consider to be the coccyx bone, just as we believe the bone that we were saying is made of six bones was called the sacrum by those who trained

---

      lumbar process until I undertook the complete dissection of an ape at Bologna for Giovanni Andrea Bianchi, and assembled its bones together with those of a human skeleton."

[68] The human sacrum is more typically composed of five bones, though it may vary from four to seven vertebrae. Vesalius admits the possibility of five, but the *Fabrica* illustrated a six-bone sacrum.

their boys at home in the dissection of bodies as well as learning the elements or letters.

The joints of the four ossicles of the coccyx are like the joints of the vertebral bodies, and a cartilaginous ligament is present here much more obviously than in the joining of the bones of the sacrum which is completely lacking in any movement. The lowest ossicle ends in cartilage and is pierced by no foramen, as no ossicle of the coccyx is seen that is pierced. It does not admit the dorsal medulla or provide a path for nerves originating from it. The sacrum is carved out throughout its length for the dorsal medulla no differently than the vertebrae, and from its upper surface beneath the vertebra resting upon it the sacrum transmits nerves ✳ in the same way we know those nerves exit from the lumbar vertebrae to the sides. In the remaining structure of the bones of the sacrum, foramina are carved both to the front and behind for the exit of nerves; there are six apiece on the anterior and posterior sides of the sacrum, so common to the two bones attached to each other that the upper bone bears the larger portion of the foramen carved into itself, unless sometimes the sacrum is formed of only five bones: in that case another joint of the first bone of the coccyx with the sacrum arises, from which a path is provided only in the posterior surface for the last nerves from the sides.

The sacrum of the dog and the ape, which can be compared in shape to the sacrum of man, is made of only three bones, which make the same attachment to each other as in the human. The inferior side of the third bone imitates the lower region of the lumbar vertebra just as the upper side of the first bone resembles the upper region of the same vertebra. The ossicles follow the lower side of the third bone and are quite similar in shape to lumbar vertebrae in nearly all their processes, especially in the attachment of the bodies and what we call the ascending and descending processes, as well as in the foramina made for the dorsal medulla and the nerves growing out of it. These ossicles number more than three in dogs and caudate apes; they are then followed by the solid and imperforate ossicles of

51

which the tail is constructed. It is different in the non-caudate apes, to which Galen provides three ossicles of the type that come immediately after the sacrum, but he does not dignify them with the name of coccyx. If therefore one compares his descriptions of apes with the fabric of humans, many things will come to light that will seem correctly written about apes but not agreeing with humans. It will immediately be concluded in other ways that Galen did not know about the six foramina made for nerves and at least three ossicles in man, and in addition ＊ he did not faithfully describe the construction of the bones of which we grant he had knowledge. Yet we do say that Galen thoroughly understood the true account of the ape. I would also like to say something about the books *De anatomicis administrationibus* and *De ossibus*. In *De usu partium*, he counted only four of the nine bones in man, having no awareness of the coccyx either in humans or in apes.

After this comes the pectoral bone [sternum], which in apes and dogs is made of seven bones that are quite clearly about the same in shape and size, not much wider than thick. However, the human pectoral bone is extremely wide at the top and much thicker there than elsewhere and quite different in shape. Close to its end it is also wide, but not thick; immediately afterward it narrows into cartilage [*processus xiphoideus*], not unlike the point of a sword. Halfway along its length, and chiefly on its upper side, the pectoral bone in man is narrower, and for that reason the ancients conveniently handed it down that it was lunate at its sides. Thus there is a very large distinction between the human bone and the ape's, and you will not find seven bones in the human pectoral bone. Whatever their number is, they are attached by a much different type of connection than in apes. In them, the articulation of all the bones is the same, but in man the articulation is extremely clear where the first bone is attached to the one beneath; the connection of the other bones is hidden on maturity and quite obscure in children also, though quite different from apes in the type, shape, and length of the bones. When the pointed cartilage of man

52

is not only brought forth between the attachment of the cartilage of the seventh rib on one side to the cartilage of the seventh rib on the other side, but also takes its beginning like that of a particular bone, this bone is so distinct from those above it that in humans of middle age there are three breast bones; and if it happens that the second bone ✳ seems in children to be constructed of many that are attached by symphysis, that articulation is not comparable in apes.

53

I have already warned how much negligence occurs in the books of Galen if we have recourse to newborns and how they purge themselves, though his description squares perfectly with quadrupeds.[69]

I have also looked at the somewhat bony cartilage of the true ribs in the very old; nothing bony appears in the cartilages of people of middle age as it does in dogs, apes, and sheep which have not yet attained their full growth, where bony and friable cartilages, and some only surrounded by cartilage as by a crust occur. So it is not surprising that Galen described bony rib cartilages of the upper ribs of animals but not of humans.[70]

By the same reasoning, if Galen had been with the Roman doctors when they cut up the German soldier and inspected his heart as he did that of an elephant, which he mentions at such length, he would certainly not have argued at such length that the heart has a bone, or at least he would have noted the difference here.[71] I should like it ascertained how evident the difference is when the heart of a

---

[69] Vesalius is recalling his critique of Galen in *Fabrica* V ch. 17 regarding the second fetal wrapping or allantois and the urachus, the urinary canal of the fetus that after birth becomes the median umbilical ligament. It is described by Galen in the 15th book of *De usu partium* 4.231.7 ff.–240.12, tr. May 1968, 664–69. Vesalius would rewrite his chapter on the fetal wrappings in the 1555 *Fabrica*, with special attention to the urachus.

[70] See Ch. 19 of Book I of the *Fabrica* under the heading "Cartilaginous Substance," where Vesalius distances himself from Galen's description of rib cartilage in humans.

[71] The matter is discussed near the beginning of Ch. 20 of Book I of the *Fabrica*. See Galen's *De anat. adm.* book 7 (§618–20) "The bone in the heart, which people think is present only in large animals and not in all of them, is there in others too, yet

decrepit ox or rather a stag is available together with a human heart; then nothing would prevent finding out whether the bony substance in a deer's heart and the ossicles themselves differ from the human. But I should be unwilling here to complain against anyone, if he had observed a substance in the heart of a decrepit old man that he could reasonably compare to a hard cartilage, as it is surprising how much softer the total constitution of man is than the dry animals just mentioned, and how long man exhibits soft cartilages and joints made by symphysis in which cartilage plays a part.

When we make an examination of the scapula and its processes and it is known to us with the greatest certitude that its higher process [acromion] to which the clavicle is articulated has been called the *summus humerus* by the leading anatomists, is it not clearer than light ⁎ that Galen, misled by what Hippocrates said, in the joint of the clavicle with that process ascribes a third bone to humans that apes lack, contrary to the fact of the matter, and which had been called the *summus humerus* by Hippocrates?[72] I make no mention of many conflicting opinions of Galen besides this joint and its process. For when Hippocrates said that the *summus humerus* was given only to humans, having conjectured from cattle, dogs, sheep, pigs, hares, horses, and quadrupeds of that sort which occur commonly, he no doubt meant the upper process of the scapula to which the clavicle is attached, since it is quite certain that those animals lack that process as well as a clavicle.[73]

---

sometimes not quite as a bone but rather as a cartilage. ... An elephant of the largest size was lately killed in Rome . ... When the heart was removed by Caesar's cooks, I sent one of my colleagues, experienced in such things, to beg the cooks to allow him to extract the bone from it. This was done and I have it to this day." (tr. Singer 1956, 186–7).

[72] *Sic*: As they wrote in Greek, the Hippocratics called it the acromion: *De articulis* 2.8, 3.21, etc. On the ambiguous use of this term in antiquity to refer to the process we call the acromion or to an abnormal sesamoid bone concealed in the acromio-clavicular articulation, see May 1968, 612 n. 51.

[73] Vesalius alludes here to a section titled "A third bone enumerated by Galen in the joint of the acromion with the clavicle" in Chapter 21 of Book I of the *Fabrica*.

But when Galen observed that process in his apes (no doubt because they, like man, have clavicles) he held Hippocrates in such veneration that he imagined a third bone exists in humans in the place mentioned, thinking that Hippocrates in his account had compared man to the ape, as he had to other quadrupeds that are served at the table. If this had been done by Galen, I would not have searched so long in vain for the third bone in man, nor for the same reason would the opportunity have presented itself in this part for me to conclude that the father of anatomy had not examined the cadavers of humans, though a truer description of the humerus along its length (for he does not always agree with himself) would likewise be attested in Galen, since what he wrote is applicable more to the dog than to man.

Similarly, Galen shows that he observed the carpal bone [*os pisiforme*] which I count fourth in the upper order, and which is called by Galen the upright and cartilaginous bone,[74] in the ape but not in man. He ascribes to it alone that it is the bulwark of a certain nerve [*ramus palmaris nervi ulnaris*] that curves from the outer part of the arm to the inner. I am not unaware of that nerve, but in humans it does not contact the bone itself, and much less does it twist along it.[75] *

Instead, it passes a little below the middle of the forearm from its inner side to the outer before sending out two offshoots to the little finger and the ring finger and one to the middle finger on their outer side. It is a branch of the nerve that I identify as the fifth [*n. ulnaris*] of the nerves entering the upper arm.

I am now immoderate in enumerating the ways in which man differs from the apes, so that I may show that human bones were

---

[74] Galen's πρόμηκες ὀστοῦν (*De usu partium* 3.131.13, 134.3.) and χονδρῶδες ὀστοῦν (*De anat. adm.* 271.6, *De usu partium* 3.135.8, 136.1), described in Book I Ch. 25 of the *Fabrica*. The passage in *De usu partium* that Vesalius criticizes here is 3.134.3 ff., tr. May 1968, 137 f.

[75] Vesalius makes this point in *Fabrica* I ch. 25, p. 118.

unknown to Galen. It sufficed Galen to consult himself in the place where he hands down the book he wrote *De ossibus*, when he has diligently inspected one by one what each bone is and what it is like, especially in those dried-out human cadavers, but if not that certainly in apes. In that account he fully indicates he has absolutely decided that everything is the same in apes and humans and that his descriptions at least fit apes if not humans. I shall accordingly set the bones aside and deal with what has now been propounded about a number of differences in other parts of the body.

Conclusions drawn about the fat, muscles, and ligaments, whereby it is concluded that Galen did not describe the human fabric

Here a large distinction between the ape and man immediately comes to mind in the arrangement of fat, which accumulates in man as in the pig in large quantities between the skin and the membrane that covers the entire body like a skin among the other membranes of the body and is commonly called fleshy. The thickness of this fat, which easily exceeds the width of a palm especially around the buttocks in women of good appearance, varies considerably both in the regions of one body and in the bodies of the obese and those consumed by wasting. I have not seen any body to have so wasted away that I did not still find some fat and some familiar substance like a glandular and fibrous fat. Today, however, I know that nobody is so slightly trained in human dissections that he has not observed a quantity and thickness of fat between the skin and the aforementioned membrane; he will also have dissected either apes or dogs, or been present at the butcher's when cattle, goats, or sheep are skinned or when hunters skin deer or hares; * without doubt he has not failed to see that there is no fat ever in these animals between the skin and the fleshy membrane, and without the intervention of fat the membrane is coterminous with the skin. Since this is so, where, I ask, in his anatomical

56

books does Galen mention the fat which we see in such quantity beneath human skin?

Though perhaps one may have decided to argue that that mass was not unknown to Galen because he made no mention of it, we must listen to what Galen himself said often, especially in the third book of *De anatomicis administrationibus* where he explains at length the artifice by which the nerves and veins running between the skin and the fleshy membrane should be examined. There he carefully sets forth the technique for inserting a knife in such a way that you do not damage the membrane along with the skin and do not remove both together from the underlying parts as butchers do.[76] In the simian or the dog it is no easy task, as he writes there and in the first book of *De anatomicis administrationibus*, to do this skillfully, and he was right with respect to simians. But no one is so witless that he would remove the skin along with the fat and the membrane in a human instead of the skin by itself. Neither in man nor in the pig does one need to be warned so much not to damage the skin with the membrane on the arms and legs, and much less still on the chest, abdomen, and the sides of the thorax when the skin alone is to be removed from it. But it is clearer than day that Galen spoke rightly about his simians and dogs and bypassed the construction of man in that part. This is to say nothing of the crowds that are stirred up in the schools when it happens that there is a silly dispute about the fleshy membrane because people navigating from books alone[77] assert with Galen that this membrane is next to the skin and that the fat in man lies underneath the membrane: those who have not come to that point of stupidity argue that all of the abdominal muscles together lie

---

[76] See *De anat. admin.* 2.348.4, where Galen instructs dissectors of the hand to remove the skin but not "the membrane beneath, through which the nutrient veins reach it." (tr. Singer 1956, 63). Vesalius is referring to his criticism of Galen in Book II ch. 5 of the *Fabrica* in the section where he describes the "fleshy rag" or *panniculus carnosus* of the abdomen, part of the fleshy membrane or fascia that in humans is separated from the skin by a layer of fat.

[77] Vesalius used this metaphor three times in the *Fabrica*. The metaphor of navigating a ship at sea by nothing more than a manual is from an adage cited by Galen, who

under the membrane, or as is read in the translators of the Arabs, they have the muscles instead of the fleshy rag.

57

I believe you still ✳ remember how many trifles the panniculus carnosus of Mondino[78] stirs up in the schools. To this should be added the difference between man, ape, and dog regarding the fleshy membrane occurring at the sides of the thorax. In those animals the membrane there is nourished by fleshy fibers and is made so muscular that Galen makes a special muscle of it, responsible for motions of the upper arm. But man has nothing in that part in common with cattle, sheep, and the animals just mentioned. It should therefore not appear surprising that I say man does not have this muscle, although with Galen I observe that fleshy place in simians and dogs.

For the same reason, I find in caudate apes the muscle that Galen said passes upward from the area of the breasts to the joint of the humerus and moves the arm to the false ribs, and I give a lengthy account of it in my book to elucidate certain passages in Galen. But since it does not occur in humans, it is another reason why I pretend not to notice that Galen dissected apes, not people.[79]

Like that muscle is the one[80] that takes a fleshy beginning in apes from the occipital bone, and with a width equal to its beginning like a muscle that is not too wide descends all fleshy the whole distance obliquely somewhat forward until it reaches the upper angle of the base of the scapula and as it ends here in a wide tendon is inserted

---

compared the study of anatomy from anatomical books to navigation out of a book. See Galen *De compositione medicamentorum per genera* 13.605.1–4.

[78] Mondino de' Liuzzi (c. 1270–1326), professor of medicine at Bologna and author of the influential *Anatomia* (ca. 1316) which was the standard manual of dissection for two centuries and the subject of a much longer 1521 commentary by Berengario da Carpi.

[79] See the section titled "The muscle adducting the arm to the chest, which is prominent in the ape" in chapter 23 of Bk. II of the *Fabrica*.

[80] See the section titled "The muscle raising the scapula that is found in the ape" in the same chapter of the *Fabrica*. It is probably the *m. rhomboideus, pars capitis*, not found in man, described in Galen's *De anat. adm.* 2.450.7ff. (Singer 1956, 107).

there on the inner surface of the scapula where we have established in its dorsal side the base of its spine. That is much different than in man, who lacks that muscle altogether, no less than he lacks another[81] that is elegantly viewed in simians by Galen, that originates with a fleshy beginning from the transverse process of the first cervical vertebra and then from the transverse processes of the third and fourth vertebrae, and becomes wider than the aforesaid muscle; he wrote that it is inserted into the scapular spine at the point where it produces the acromion, and is established as if directly opposite the neck of the scapula. I do not miss this muscle in simians, ✳ but I know very well that humans do not have it.

58

I would also like it to be noted that a muscle common to simians and humans, which also arises from the transverse processes of the upper cervical vertebrae and is inserted in the upper angle of the base of the scapula, is stronger and larger in man than in simians.[82] It should not go unmentioned here at what length Galen describes the upper part of this muscle[83] and nearly forgets himself arguing about it with Lycus; I usually compare it to the hoods of monks and call it the second of the muscles moving the scapula. But after describing many muscles, Galen devotes scarcely a word to the lower part of this muscle. One should contrary to the usual opinion be convinced by this that the muscle was not dissected in humans but rather in dogs.

In addition, Galen never saw the muscle in man that he said in two places takes its beginning from the hyoid bone and is implanted in the scapula, and that the scapula is raised by it (he was deluded in

---

[81] See the section titled "A fourth muscle which apes have but humans do not" in the aforementioned chapter of the *Fabrica*. It may be identified as the simian *m. atlantoscapularis anterior* which Galen mentions in De anat. adm.2.469.2ff. (Singer 1956, pp. 116f.)

[82] See the section headed "Third of the muscles moving the human scapula" in chapter 26 of the second book of the *Fabrica*. This is the *m. levator scapulae.*

[83] Vesalius is now talking about the trapezius, described in the section of Bk. II ch. 26 entitled "Second of the muscles moving the scapula." Galen's description is in De anat. adm.; The passage cited by Vesalius is in the sixth chapter of Bk. 4 (2.449.3ff.,

the function of that muscle);[84] otherwise he would not have written that it is fleshy throughout its course, as can be seen in simians. He also would not have said that no muscle whatever in the entire body looks like the one[85] to which we say the function is entrusted of drawing the lower maxilla downward: it is formed as if with two bellies and becomes tendinous halfway in its course, taking the shape of a rounded tendon. Like the muscle from the scapula inserted in the hyoid bone, it appears sinewy in man where it enters the muscle that originates from the clavicle and the sternum and is implanted in the occipital bone, serving motions of the head.

This muscle that we happened to mention shows that Galen by no means dissected muscles moving the head in man, as will be obvious to anyone who has compared a caudate ape with man here and has learned to apply Galen's description, which is more accurate ✳ in different places, unlike this one, to the ape. There, the origins of the aforementioned muscle are more distinct than in man, and another muscle is extended to it and attached to it first at its insertion much differently than we (together with the anatomists who preceded Galen) discovered in man: we determined a single muscle here, though in apes it would easily be possible to increase the number.

What I now have to say, however, will demonstrate more clearly still that the second pair of muscles moving the head,[86] as their

---

Singer 1956, 107) where Lycus is criticized for saying that the trapezius draws the head to the shoulder rather than the opposite.

[84] Vesalius makes this argument in *Fabrica* Bk. II ch. 17 in the section headed "The fourth pair [of hyoid muscles] does not assist motions of the scapulae." In Book 13 of *De usu partium* Galen had written "the slender muscle [*m. omohyoideus*] arising from the lambdoidal [hyoid] bone draws [the scapula] forward; for this muscle too is inserted into the bone of the shoulder blade near the acromium." (4.140.4–7, tr. May 1968, 618). He made a similar statement in Book 7: "the muscles that extend [from the hyoid bone] to the shoulder blades give them a motion toward the neck." (3.592.18–593–1, tr. May 1968, 375 and n. 69).

[85] The digastric muscle.

[86] These are described in Bk. II ch. 28 of the *Fabrica* in the section titled "The second pair is quite various."

description in Galen argues so clearly, is the one he dissected in the dog and the ape, since it matches those animals but by no means man. In that pair he shows so many muscles that are quite different in form, in many bellies, tendons, and multiple ways of coming together, whose description he cannot accomplish except in a lengthy account, from which it will be possible to refrain because I believe it is all too clear to people who undertake an anatomy that the muscles moving the human head were not taken into consideration by Galen.

Again, if we have noticed that the human thorax is shorter than the thorax of quadrupeds, it will be no surprise that a different set of muscles moving the thorax is found in dogs and caudate apes than in humans, and beyond doubt it is from this that the strength of my proof is evident. Galen counts the muscle extending along the

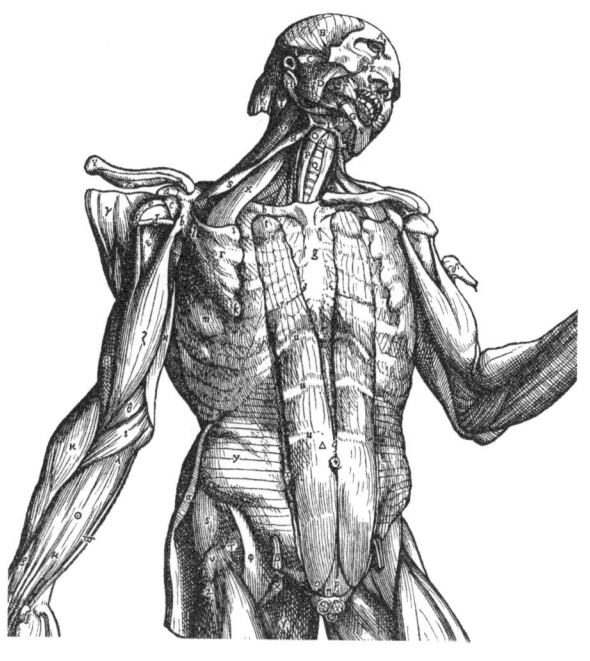

Detail of the fifth écorché in the series at the beginning of Book 5 of the Fabrica, showing the upward continuation of the rectus abdominis (r to t) as it would appear, say, in the dog.

anterior part of the thorax, which I find in dogs and caudate apes, as part of the rectus abdominis muscle. For where this muscle ends above the rib cartilages in humans, at that point in those animals it puts forth a kind of long tendon extended along the rib cartilages beside the sternum and next to the cartilage of the second rib, augmented by a fleshy protuberance like a flat stomach muscle. Because none of this appears in man I cannot deny faith in my eyes and ascribe the muscle to man solely on the authority of Galen.[87] ✳

60   So I do not attribute to man the muscle [*m. scalenus longus*] that was directly apparent to Galen as it is to me in dogs and simians, taking its beginning from several transverse processes of the cervical vertebrae and serving motions of the upper ribs; it runs along the anterior surface of the muscle [*m. serratus anterior*] that originates from the scapula and is inserted like a hand on the eight upper ribs. I count the latter the second of the muscles moving the thorax.

I know that I have made my account of the lumbar fleshes (to use the word of other anatomists) quite different from Galen's, and that a large difference presents itself here between man and the apes; but because Galen's descriptions in this place have seemed to me quite mixed up and difficult, but do not everywhere accord with dogs and apes, I can take no argument for myself by which it may be proved that Galen did not dissect humans.[88] However, no one can be in doubt that Galen's descriptions have been matched to dogs much more than humans, particularly because of the muscle of which they have so much. Similarly, in the account of the ligaments of the spine you will notice that human ligaments were not dissected by Galen – or any of those that he says extend along the length of the back to the vertebral

---

[87] This muscle is described in *Fabrica* II ch. 35 in the section titled "The fifth muscle in dogs, missing in the human thorax." It is illustrated in Vesalius' Fifth Table of Muscles as if it existed in man. The 1555 edition of the *Fabrica* adds to the figure legend "The wide tendon and this fleshy part are the muscle that Galen counts the fifth of those moving the thorax, but it is not to be seen in humans as it is in caudate apes and dogs. We have nevertheless drawn it here so that Galen can be understood, because elsewhere this part of the chest would be like the chests in the next two tables, without these muscles."

[88] Vesalius refers here to his remarks in ch. 38 of Bk. II of the *Fabrica* in the section titled "Ninth and tenth" [of the muscles moving the back]: "How clearly and truthfully Galen described these muscles or fleshes of the loins (to use the translator's word), and whether his account fits the ape or dog (which differs considerably from man in the muscles here) will be for you to judge. It does not quite square with the ape, much less with man, as you will understand perfectly (if you are a student of Galen, as surely we all should be) when you perform some dissections and compare what we say with his oracles."

spines; they are never found in humans as they are in cattle and sheep, unless you have the best reason for finding them.

If we have observed the tendon [*m. palmaris longus, tendo*][89] hidden in the hand, we would not dare to assert that it extends everywhere beneath the hairless skin of the hand since the fleshy mass occupying the first joint of the thumb known as the Venus mount by palm readers, and the mass that stands at the outer side of the palm below the little finger which they refer to as the Moon, and likewise the sides of the fingers, are not covered by that tendon. However, when we study the hand of the simian and observe a hairless region there, the account in Galen is understood to be more consonant with the truth. It is not unlike what Galen stated when he was more experienced in dissections ✳ and did not so much invent the fabric of the body and 61 write from the opinions of others or his own imagination, that the tendon to which is entrusted the task of flexing the third joint of the thumb is thinner than those that flex the third joints of the remaining fingers. That is because we see that a more or less special muscle in man puts forth an extremely strong tendon to the third bone of the thumb; this muscle is not at all inferior to the others that go to the third bone of the fingers. But it is not strange that such an idea was passed on to posterity by Galen, since we know that the simian thumb is much inferior to the human thumb in both strength and size, and should therefore be content with a thinner muscle. That he made his examination of this muscle and its tendon in non-caudate simians, I gather from the following: that in caudate simians and others that have a small, weak thumb, this muscle would be missing, and its functions would be performed by a portion of the small tendon that flexes the third bone of their middle finger. From the middle of the palm, a small portion of tendon originates from the tendon that branches transversely to the thumb. I think that Galen would at some time have seen something

---

[89] Vesalius devoted all of Bk. II ch. 41 to the palmaris longus and its tendon, and refuted Galen's description of the tendon in *De anat. adm.* and *De usu partium.*

71

like that, since in *De usu partium* he ignored the five muscles that cause flexion of the first and second bones of the thumb and wished to make up a reason why he would show that the thumb is flexed over the remaining fingers toward the middle of the palm. He stated that a certain muscle was divided into five tendons before it passed through the wrist, four of which made three equal angles and went to the four fingers; the fifth tendon, along with the one that goes to the middle finger, was gathered together and taken to the middle of the palm, and from there, like the rein with which a charioteer drives a horse, it was extended from a ring transversely to the thumb.[90]

But now it is not my wish to run through any of the false descriptions, contradictory opinions, or an action or function unknown to Galen, but only those things that force me to say he did not dissect a human. ✳

62   To this category also belong the muscles that extend the forearm, which Galen described in the first book of *De anatomicis administrationibus* (in *De usu partium* everything about the muscles moving the forearm is absurdly fictitious) as they are in the ape. It is therefore no surprise that he counted three muscles among the extensors (unlike what we see in man).[91] While one originates in man from the humerus and another from the lower rib of the scapula, both are joined and come together before the mid-length of the humerus as if the beginnings should be thought of as belonging to a single muscle. Then, since I wrote that the beginning that originates from the humerus grows in its fleshy portion as it descends, it becomes quite

[90]   These remarks summarize and extend criticisms of Galen made in ch. 43 of Bk. II of the *Fabrica*, on the finger muscles. See especially the section titled "Many things are set forth here contrary to the views of Galen."

[91]   The 1543 *Fabrica* did in fact identify the three extensors of the forearm that modern anatomy counts as a single muscle, the *triceps brachii*, but he changed his mind in the 1555 *Fabrica*. See n. 12 in my edition of *Fabrica* Bk. II, Second Table of Muscles, where O is the *caput laterale* and P is the *caput longum*. In the legend for Q he mentions the *caput mediale* as the third muscle, but changes that to second in the 1555 edition. The same change is made at X in the 12th Table of Muscles.

difficult to consider it as a separate muscle. But apes have three distinct muscles, and Galen's inconsistent accounts show no muscles except for a single quite thin muscle that I have said in my book is borne from the scapula to the ulna in caudate apes.

I do not remember clearly whether the beginning of the anterior of the muscles flexing the forearm [*m. biceps brachii*], which arises from the inner process of the scapula, is more slender in simians than the one that takes its origin from the top of the neck of the scapula, as Galen liked it.[92] However, in man it appears quite different, as the inner beginning is quite wide, partly sinewy and in part fleshy, while its outer beginning resembles only a rounded tendon. Similarly, the inner origin is also sinewy perhaps only in simians, presenting the appearance of a tendon because the inner process of the scapula is extended less in them than in humans. For this reason Galen is seen to be more readily deceived in that muscle, which because it performs only motions of the scapula he wrongly made the adductor of the upper arm to the chest. In simians, this muscle at its insertion faces somewhat into the ligament of the joint of the scapula with the humerus, ∗ since it was able to make such a small insertion into the inner process of the scapula (differently than in man); the complete insertion extends into the otherwise large inner process of the scapula.[93]

The first of the muscles moving the tibia [*m. sartorius*] originates from the anterior side of the epiphysis of the iliac bone and is taken on an oblique path ultimately to the tibia; in humans it is inserted by a round tendon, but in simians, as Galen also testifies, by

63

---

[92] Vesalius faults Galen's description of the muscles of the upper arm in the section of Bk. II ch. 23 titled "The muscle adducting the arm to the chest, which is prominent in the ape."

[93] This critique of Galen goes beyond any found in the *Fabrica* (Bk. II ch. 46 on the flexors and extensors of the radius), where Vesalius had written "I shall reveal all of this in my annotations to the Anatomical Works of Galen, which I have already well begun and shall at some time publish separately or together with the books of Galen much better corrected than formerly."

a wider tendon.[94] However, it is clear not from this muscle alone but from many other movers of the tibia that Galen left posterity with a description of the muscles in simians, since he ascribes width to all the tendons inserted on the front of the tibia (as we definitely see in simians) while in man they are rounded and more compact. This difference is seen to be more pronounced in dissections than one would believe from books alone. Now too a muscle quite familiar to Galen, and called wide by him, shows no less a distinction; we call it the fourth of the muscles moving the tibia [*m. biceps femoris, caput longum*], and it has an appearance in man far different from that in the simian. This fourth muscle in the simian issues from the lowest surface of the hip bone, as Galen rightly said (though he errs in telling the series and order of heads originating therefrom) with a strong beginning, which early in its progress forms a larger, wide muscle that is simple throughout its course, moving along the outer head of the femur to the tibia and the seat of its insertion there. In man, however, the beginning of the fourth muscle is entirely sinewy, and a little beneath its origin it first becomes fleshy, presenting the shape of a completely formed muscle in the form of a mouse or lizard, but not of a wide muscle. When it has more or less passed the halfway point down the femur it becomes thinner and appears notably sinewy on its outer side, being about to end in a tendon. But on its inner side, where it still preserves the nature of flesh, * a fleshy portion [*m. biceps femoris, caput breve*] to which the femur provides a beginning attaches to it just as if a peculiar muscle were increasing it and both together make up a thickened muscle which is eventually very strongly inserted in the apex of the fibula.

A portion of the muscle identified as the fifth of the muscles moving the femur [*m. adductor magnus*] which is implanted in man

---

[94] In *Fabrica* Bk. II ch. 53 Vesalius called attention to this error in Galen. A marginal note (p. 331, misnumbered 231) cites Bk. 2 of *De anatomicis administrationibus*, where Galen called the tendon of the sartorius muscle "a flat tendon, somewhat fleshy," τένων πλατὺς ἠρέμα σαρκώδης, (2.293.13 f., tr. Singer 1956, 37).

by a round tendon into the inner head of the femur, makes a fleshy insertion in simians. It is therefore no surprise that Galen puts it in his writings no differently than if it were the same in man as well.[95]

More or less similar to this place is the difference observed in the tendon [*aponeurosis plantaris*] hidden under the sole of the foot. In simians, we are able to see the muscle that originates in the outer head of the femur; after it has reached halfway down the tibia it ends in a thin tendon that is carried through its own groove carved in the back of the heel bone and enters the sole, hiding there as is seen in the hand. In the opinions of Galen, the ape is the same in this part, except that no portion of this tendon is implanted in the heel bone; rather, it is slightly linked to the heel by means of the transverse ligament. Galen, however, taught that this tendon is split in two, with one portion attached to the heel bone while the other enters the sole of the foot. In man, on the other hand, we do not detect the muscle brought out into the hidden tendon, though we do not miss the ligament growing on the muscle [*m. flexor digitorum brevis*] that we name the flexor of the second bone of each of the four toes. It is perfectly clear that it performs the functions of such a tendon from the fact that the muscle split into four tendons before they enter the transverse ligaments of the toes no longer bears that ligament attached to it. But as it leaves the muscle there, it is separated into five processes spread out in the lower region of the five digits of the foot. It is not possible that anyone would believe that the muscle I have named the third of those moving the foot [*m. plantaris*] is applicable to Galen's account, since before it has passed ✳ a noteworthy distance beyond 65 the knee joint it ceases to be fleshy and is attached to the inside of the heel at about the middle of the distance one could measure from the tibia to the end of the heel. Moreover, it is not so great a task for that

---

[95] The portion of the adductor magnus described here is the hamstring portion, which inserts into the adductor tubercle of the femur. This criticism of Galen is another not found in the *Fabrica*, and could therefore be one of the comments Vesalius had intended for his projected edition of Galen. See n. 93 above.

ligament to be compared to the aforementioned ligament peculiar to humans.[96]

At this time the tendon of the first two muscles moving the foot [*m. gastrocnemius, caput mediale* and *caput laterale*] should be carefully examined. These originate from the heads of the femur and create the more protuberant region of the calf; they end together below the middle of the tibia in a strong, wide tendon that runs downward in caudate apes and as it becomes gradually narrower is attached to no muscle; it is inserted unmixed into the heel bone, as Galen's account nicely explains. It is much different in humans,[97] in whom the tendon passes far beyond the heel bone, joins completely with the tendon of the fourth muscle moving the foot [*m. soleus*] (which also forms the calf together with the two muscles just mentioned), and is so attached that you could scarcely separate it in one piece as far as the heel bone.

To this distinction is added the one that a person would rightly judge worth considering in the tendon of the fourth muscle just mentioned. Galen contends that this muscle, originating from the tibia and the fibula, does not end in a complete, fleshless tendon, but makes its insertion still fleshy in the heel bone. This is how that muscle is inserted in simians, as I know. But it is otherwise in man: for a long interval before this muscle is inserted in the heel, it ends in a perfect tendon that in no way resembles the nature of flesh; increased by the tendon of the first two muscles, it goes to the heel. Homer was certainly not ignorant of this when he describes the course of the rope by which Hector, tied to the chariot of Achilles, was dragged around the walls of Troy.[98]

We should also not pass over here the visually elegant difference between the muscles that lie beneath those just mentioned in

---

[96] The antecedent to this paragraph is chapter 58 of Bk. II of the *Fabrica*, "On the Hidden Tendon Attached to the Skin on the Sole of the Foot."

[97] This distinction is explained at the beginning of ch. 59 of *Fabrica* II.

[98] "In both of his feet at the back he made holes by the tendons in the space between ankle and heel, and drew thongs of ox-hide through them, and fastened them to the chariot so as to let the head drag, etc." *Iliad* 22.396 ff., tr. Lattimore. This episode was also cited in *Fabrica* II ch. 59.

the posterior region of the tibia, ✳ two of which are responsible for
motions of the toes. I count them the second [*m. flexor hallucis longus*]
and third [*m. flexor digitorum longus*] of the muscles moving the toes; the
one covered by those two, lying closest to the tibia and fibula, is considered the fifth [*m. tibialis posterior*] of those that move the foot. Before
they reach the lower epiphysis of the tibia, these muscles in simians
are stripped of their fleshy substance and like other muscles that have
round tendons they end in a tendon. But in man they have a peculiar
form; when they are about to put forth the tendon they appear about
as wide as in the middle of their course, and they put forth a tendon
from only one side while the angle of the other side is still fleshy as if
it were to be regarded as the angle of a fleshy quadrangle.[99]

Still more relevant to the present topic is the difference in man of
the sixth muscle moving the foot [*m. tibialis anterior*], which is at variance with the construction of the ape and conflicts with the descriptions of Galen.[100] In man, the muscle in front of the tibia standing
ahead of all the others in this region takes its beginning where the
fibula is joined to the tibia and resembles an elegant type of muscle;
below the middle of the tibia it ends in a round tendon that is carried
by the transverse ligament at the front of the tibia next to its joint with
the talus and ends in the metatarsal bone that supports the big toe. In
caudate apes, however, that tendon not only becomes two-horned
to look like the eighth muscle moving the foot [*m. fibularis brevis*],
which is brought down along the lower epiphysis of the fibula in the
transverse ligament and inserted in the metatarsal bone that supports
the little toe – not only that, in caudate apes two quite conspicuous
muscles appear to dissectors, of which the posterior is much thinner
than the anterior and is laid over it: both are implanted in the bone
that I said admits the sixth human muscle.

---

[99]   This observation is independent of anything said in *Fabrica* II ch. 60 on the muscles
moving the toes.
[100]   This criticism of Galen does not appear in either edition of the *Fabrica*.

The muscle that causes flexion of the second joint of each of the four toes [*m. flexor digitorum brevis*] differs in man as much from the ape as the hidden tendon ✳ with the ligament attached to that muscle, except that it is conspicuous for longer tendons in apes to the degree that their foot is longer than the human.[101] It should not seem strange that we ascribe a longer foot to the ape than to man: Galen shows in a long account that it is the longest of all. But how truly

Detail from fig. 14 in Bk. II of the Fabrica, showing various flexor muscles in the sole of the foot. Λ marks a segment of the flexor hallucis longus, Vesalius' second of the muscles moving the toes; Ƶ is the flexor digitorum longus, Vesalius' third of the muscles moving the toes; κ marks the tibialis posterior, Vesalius' fifth muscle moving the toes. Θ is the flexor digitorum brevis; on the right it has been severed from its origin and hangs from its several insertions.

he says this is easily known to anyone who pays close attention and is warned by me of a contrary view. Now the big toe on the simian foot is different from ours, as is the different separation of the toes, which Galen did not fail to mention in his books since he knew it without dissection; these also should have taught him that there is a much different construction of the muscles moving the toes in humans than in apes. For although he is forgetful of himself in describing those things, you will nevertheless say that the opinion closest to the

---

[101] See *Fabrica* II ch. 60, the section titled "How the first, second, and third muscles are arranged in the apes."

truth is no doubt the one which most closely agrees with simians in large numbers differing from humans. Lest I pass over all the tendons moving the toes, I think the tendons of the second and third muscles that move them are worth considering. Previously, I have told how the muscles that occupy the back of the tibia put forth their tendons differently in humans than in simians; the series to which they belong must now be examined. The tendon of the second muscle in man moves obliquely through the bottom of the foot and goes chiefly to the big toe, inserted very strongly in its second joint. The tendon of the third muscle is carried obliquely beneath it crosswise toward the ground, or is situated with it like an X. Taking on a tiny portion from the tendon of the second, and slightly increased by it, it is divided into four tendons going to the four toes, creating flexion of their third bone. These are as much thinner than the one that goes to the big toe as the big toe surpasses the others in thickness and bulk. In caudate apes, where the tendon of the second muscle goes crosswise with the tendon of the third, ✳ the entire tendon of the second 68 muscle is altogether blended with the tendon of the third and joins with it, and so there is one tendon out of two. The tendon resulting from that juncture puts forth a slender tendon or small portion transversely to the big toe to flex the bone to which the toenail is attached, just as in the palm of the caudate ape I was saying that the tendon flexing the third bone of the middle finger gives off a small portion transversely to the thumb of the hand. The principal portion of that blending is separated into four tendons and extended to the other toes; these tendons are much thicker than the small portion going to the big toe. It should not be surprising that the different construction of the big toe creates many differences in the muscles, such as the one arising from the muscle that comes separately from the metatarsal bone supporting the second toe, is inserted in the big toe of the ape, and adducts it closest to the second toe.

Similar to that is another series belonging to the fleshy mass [*mm. extensor hallucis brevis et extensor digitorum brevis*] in man located in the upper part of the foot: it supplies one tendon to the big toe and the three toes closest to it, inclining them with the toe to the outside. In the simian big toe as in its hand there were two muscles implanted in its upper side, the thinner of which is employed in the function of the tendon that I was just now saying goes to it from the upper side of the foot. Then two quite confined muscles occupy the upper side of the foot, by which the four toes are abducted from the big toe (as Galen rightly wrote). Thus the muscles flexing the first joints of the toes in the human foot cannot so easily be summed up in a number and appear more like an inseparable fleshy mass than in simians, in whom they exhibit a more elegant arrangement, as they do in the muscles of their hands.[102]

But why does it seem strange that so many differences present themselves in great numbers in the muscles of the foot, when we observe a great difference in their motions and never discover the same construction in a variety of operations, actions, or functions? Let it suffice, therefore, that I have cited these places from the study of the muscles and ligaments in which I demonstrate that Galen did not dissect a human. ✳

69 *Several places taken from the series of veins and arteries in which it is inferred that Galen did not dissect humans*

It likewise happens in the veins and arteries that I think no one's authority will have the same weight as his argument. For if Galen, when about to write the book *De venarum arteriarumque natura* about the veins and arteries and the system by which they are distributed through the entire body, openly warns Antisthenes that he is going to write out as notes the facts about veins and arteries which he

---

[102] This complexity, as opposed to the simplicity described by Galen, is explained in *Fabrica* II ch. 60 in the section titled "A muscle situated in the top of the foot, abducting the big toe and the three next to it to the outside, numbered 16."

had himself seen in apes, with no mention made of man,[103] what conclusion are we to draw except that the course and distribution of the veins and arteries of apes is described by Galen? – the more so, as I believe that in none of the books of Galen does he himself mention any difference in which he was aware that apes differ from man in the branches of the vessels? You could perhaps make an exception of his tenth commentary on Hippocrates' *On Regimen in Acute Diseases*, on the occasion that I shall soon describe where he imagines contrary to the truth of the matter that the human azygos vein takes its beginning from the vena cava, unlike that of apes. As soon as I have set forth certain noteworthy differences out of many in which man varies from the ape and the dog in the distribution and course of the vessels, I shall add something about the azygos vein.

If we consider that throughout its course where it extends beneath the stomach or where it runs from the site of the spleen to a point beneath the liver, the large intestine has no part of the mesentery in humans, but lies entirely beneath what we call the lower membrane of the omentum in the posterior side of the stomach running from the spleen to the liver instead of in the mesentery itself, it is established that it is different from what it is in simians and still more so from what it is in dogs. It will soon also be clear that the series of branches of the portal vein is different in man than in the simian. For from the trunk of the portal vein, which is supported by the lower membrane of the omentum and extends along the main portion of the spleen, large veins are woven into the large intestine of

---

[103] In *Fabrica* III ch. 5 when describing the distribution of the left trunk of the portal vein Vesalius wrote "it is no surprise if I sometimes depart from Galen's views in this part of my account. He openly writes about the veins of monkeys, as he himself wrote to Antisthenes, and does not deal with the veins of humans." A marginal note refers to Galen's work *De venarum arteriarumque dissectione*, at the beginning of which he tells his contemporary Antisthenes of Rhodes that his synopsis of veins and arteries is based upon the body of the monkey. Vesalius had edited Antonio Fortolo's Latin translation of this book for the 1541–42 Giunta edition: see Cushing 1962, 66; O'Malley 1964, 106.

man throughout the section just described, which it is no surprise that Galen did [not] describe, since he handed on to posterity a description of simians and dogs rather than of humans.

I will not discuss how two veins ✳ interweave the fundus of the stomach in humans and attach to the upper membrane of the omentum; one of these enters the stomach from the right side, the other from the left. Both end at the middle of the stomach. However, in dogs a single vein going to the stomach from the left side, which runs along its entire fundus as far as its lower orifice, is inserted from the lowest vein of the spleen.

Similar to this difference is the one that arises from the large series of other offshoots growing from other veins that are now close to the spleen and are disseminated throughout its body. Though Galen does not mention more than two offshoots, they are definitely many and various in the human spleen at its connection to the stomach.

But I did not propose to mention here everything available that is scattered in the books of my *De humani corpore fabrica*, since it would suffice to mention something about most of the organs individually. It will therefore be fitting that I have already mentioned the vena cava, which principally will show (if it does nothing else) that Galen never saw a human dissection, or what one would easily have noticed from the course of the vena cava through the transverse septum. As the dog and the ape greatly surpass man in the length of the thorax, so too the course of the vena cava through the thorax is deservedly different for man and for an ape.[104] And so the vena cava of the simian or dog passes through the transverse septum and enters the cavity visible between the two membranes separating the thorax and between the wrapping of the heart and the transverse septum where there is a peculiar lobe or fiber of the lung supporting the vena cava like an extremely soft bed, holding it up more or less like a hand in this long

---

[104] This the subject of a section in *Fabrica* III ch. 7 titled "Features that do not at this point agree with Galen's views."

passage in which it is otherwise suspended and carried by no body, not without the unique design of Nature. From this space, the vena cava is contained in this way by the lung and passes through the wrapping of the heart, finally opening into the heart via an orifice much larger than its own stem. *

Furthermore, the wrapping of the heart in man is attached to the septum by its entire lower surface of considerable size, and no interval presents itself between the wrapping of the heart and the septum, in which there could be a cavity between itself and the membranes that divide up the thorax. So too there was much less reason why a peculiar lobe of the lung needed to be placed here to support the vena cava. Therefore the vena cava passes through the transverse septum together with the wrapping of the heart without the aid of any empty space or any body, for this reason differing in many ways from the course of the vena cava in apes and dogs, a distinction not noticed by Galen as he did state a difference in the beginning of the azygos vein. It is the only one, as I recall, among all that one could count beneath the skin in dissections. It should therefore by considered with exact diligence by those who argue that Galen dissected humans, even after being warned, so that those who perceive no difference there should not be in a hurry to set aside their own reason because of piety toward Galen.

It is a precept of Hippocrates that when pain in the side reaches as far as the clavicle, the inner vein of the forearm should be opened; but when the pain stands lower down, even below the septum, a purging medication should be administered.[105] Galen understood this precept just as if Hippocrates * had recommended bloodletting in tumors that affected the four upper ribs; but when in some cases the trouble lay in the eight lower ribs, a purgative drug should be administered. It is no different than if Hippocrates had believed those ribs

[105] Vesalius' statement here is anachronistic, coming from medieval venesection lore unattested in the Hippocratic writings, or attributed to Hippocrates by Galen or Pollux. For what the Hippocratics actually wrote about bloodletting, see Brain 1986, 112–21.

were too distant from the veins to be cut, or because he feared that the blood close to the inflammation would be taken through the heart whenever the brachial vein would be opened for inflammations of the lower ribs. This follows the reasoning on which we know Avicenna's *Canon* was based, that we should not take poisonous blood through worthier parts, as if Hippocrates had prohibited veins in the forearm to be opened for inflammations of the liver, stomach, and spleen. You know that I have explained elsewhere how I understand the place in Hippocrates, and that I have also showed from the aforementioned opinion of Galen about that place that he believed that the divine Hippocrates, who was practiced in human dissections, had found a different series of veins in his apes than he had himself, which persuaded Hippocrates to prefer a purgative drug to venesection.

For that reason, Galen imagined a different origin for the azygos vein in humans than in apes, and because of the oracle of Hippocrates he became so agitated that he seems to have pursued quite contradictory thoughts about the beginning of the azygos vein. In his recently mentioned commentary on the second book of *De victus ratione*,[106] Galen disagrees with himself three times.[107] When about to describe the system of the vena cava in the thorax, he states everywhere that the vein distributing offshoots to the eight lower ribs takes its beginning from the vena cava before it reaches the heart. After that, he writes that in some animals it originates from the vena cava above the heart, whereas in man he writes that it is put forth from the part of the vena cava where it reaches the right auricle of the heart. Then, unmindful of himself, he says a little later that the lower ribs of man are nourished by the vein that begins from the vena cava beneath the heart. This opinion resembles the one in Book Six of *On the Opinions of Hippocrates and Plato* where we are told that the wrapping of the

---

[106] Hippocrates' *On Regimen in Acute Diseases*. Galen's commentary on *De victus ratione* is *In Hippocratis de victu acutorum commentaria*.

[107] For these observations, cf. the section in *Fabrica* III ch. 7 titled "Galen's views about the origin of the unpaired vein."

heart together with all the membranes partitioning the thorax and covering the lung take nourishment from the vena cava before this vessel is taken to the heart. I have no doubt that when he mentions the membranes surrounding the lung in that passage, he meant the membrane surrounding the ribs. But in his book *On the Dissection of Veins and Arteries* Galen stated that the azygos vein sets out from the right auricle for the left parts of the thorax and is taken to the fifth vertebra of the thorax in certain animals, while in simians he says it is placed a little above the auricle in the parts on the right. ✱

He says more here than in his annotations on *De victus ratione*, adding that the intervals between the ribs seek nourishment from the azygos vein. He expresses the same opinion in the seventh book of *De anatomicis administrationibus*, except he states simply that the vein takes its beginning here from the vein standing next to the right ventricle of the heart – though he adds wrongly that the azygos vein is taken there to the *left* parts of the thorax.

At another point in his seventh book he teaches that this vein originates from the right ventricle of the heart,

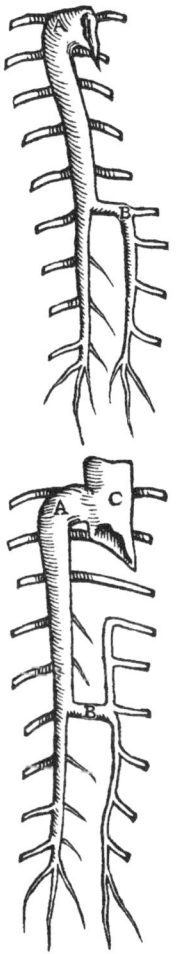

Two variations in the form of the azygos vein from Fabrica III ch. 7, where Vesalius wrote "In these two figures we have shown two arrays of the azygos vein that differ from the one shown in the complete illustration of the vena cava. Here, A marks the trunk of the azygos vein. B is its branch [v. hemiazygos] that runs to the left and is distributed into several offshoots. C in the lower figure shows part of the stem of the vena cava."

just like the coronary, although this never happens in simians. It is therefore all too clear that Galen believed that the azygos vein takes a different beginning in man and in simians. It is also no surprise that such different statements about this vein occur in his writings because he never examined the human vein. For him it originates in exactly

73

the same way as it does in apes and dogs and corresponds to theirs in all ways – unless perhaps the human vein nurtures the upper ribs more than the vein in dogs. But we have observed that all the ribs are nourished by the azygos vein in humans and receive its offshoots. Although Galen had not examined the human, he was nevertheless compelled to look into the reasons why the azygos vein could not have taken its origin beneath the heart in humans, or even directly opposite the right auricle of the heart, and that no difference could be imagined here between the ape and man. For although Nature arranged for the stem of the vena cava to attach to the spine from the lumbar vertebrae to the upper vertebrae of the thorax, or for the offshoots from it at each node as in the loins, she was unable to present them to the intermediate vertebrae or the ribs; because the transverse septum, the liver, then the heart and the vessels distributed to the lung elevated the vena cava too far from the vertebrae and did not allow it to recline, it was necessary for her to lead off an offshoot like a trunk from the vena cava from which, * while it was braced in its passage on the vertebral bodies, branches could be extended to all parts; because of the distance of the vena cava from the vertebrae it was difficult for these parts to receive twigs from it. Here Nature, as never elsewhere, did not miss the opportunity, but as soon as the vena cava passed the heart and traveled beyond its wrapping, both the processes of the great artery [aorta] alone and the trachea and the path delivering food and drink are prohibited from resting on the vertebrae and Nature brought out of the right side of the vena cava the large trunk that extends downward on the right side of the vertebrae and bestows branches on the ribs of both sides. We call it the unpaired vein [*v. azygos*] because another one paired to it, or a partner, does not originate from the vena cava.

From the course that it takes along the right side of the vertebrae at the middle and impeded on its left by the great artery [*aorta*] and the passage that delivers food and drink, it is to be inferred that it could have taken its beginning neither from the left side of the vena cava nor from its lower side, and still less from its upper side. From this it is clear that it could not have been brought forth under the heart, for there was need for another vein to be presented to that interval of the back which is located between the origin of the former and the place where it now takes its origin. In such a case it would have hung too much and been braced too little if it had originated under the heart and been extended to the vertebrae, besides the fact that in man it would then have had to begin from the vena cava when it was still passing within the wrapping of the heart. That is a proof that the azygos vein does not begin from the vena cava directly opposite the right auricle, and here the arterial vein [*truncus pulmonalis*] and the venous artery [*vena pulmonalis*] would have been in its way, and then the branches of the trachea as they first separate into the lungs beneath the base of the heart. Because of them, it would have been necessary for it to climb too high before it could approach the spine, and it is only taking its beginning; or it would have had to pass too low and have gotten its beginning under the heart besides.

Galen did not see the inner veins that hide deep in the human arm[108]

Here I would have made an end of sampling the proofs by which I am persuaded in the system of veins that Galen did not dissect humans, * had not Sylvius, while making no mention of any other particular places, been specially offended to have read in my book that Galen saw only the outer and subcutaneous veins in man without dissection. To me this is as likely as it can be. I know that there is in no series of vessels as great a difference as there is in the veins running under the skin of the forearm and the hand, so that we see few people who have exactly the same set of veins. But because I noticed that Galen's description of this set of veins in the third book of *De*

75

---

[108]   The background of the narrative that follows is in *Fabrica* III ch. 8 on the axillary and humeral veins.

anatomicis administrationibus agreed most with the upper arms, at least in the course of the major branches, I easily concluded that those

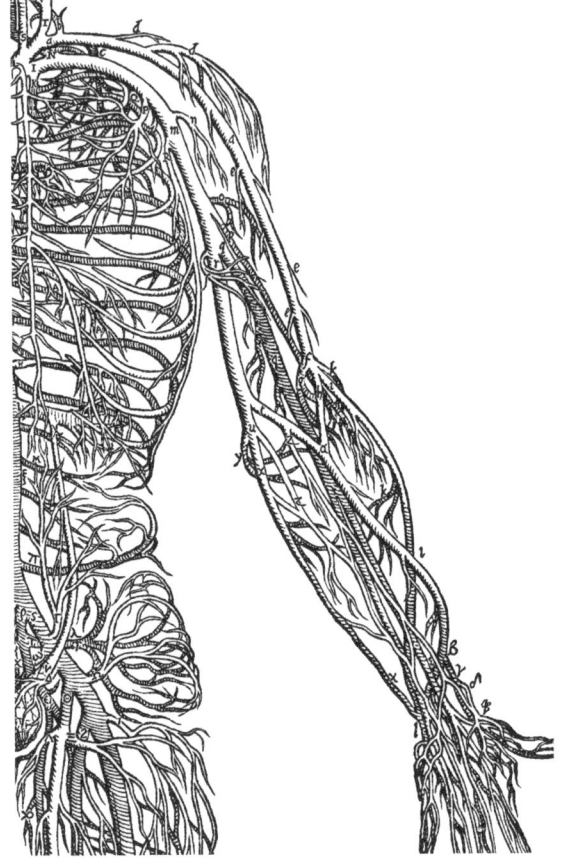

veins had been seen by him as we see them daily in venesection. But I did not believe that Galen had seen the series of branches hiding deep in the body, since his description does not for the most part agree with it.

I will not mention that we have been wrong about the origin of the humeral vein,[109] which we thought originates from the jugular much higher than the axillary,

Detail of the "Vein Man" preceding Fabrica III Ch. 6. The humeral or cephalic vein is marked a; the axillary or basilic vein is m, r, and u. Greek α is the distal end of the median cubital vein (t) discussed below.

to which the division of the vena cava in the throat gives its origin, and which is twisted in one way or another along the clavicle; because we trusted in books, we do not see that the humeral vein near the elbow joint in some way inserts a noteworthy branch more deeply, and with

---

[109] *Vena humeraria*, a Galenic vein: "The veins of the arm are two, the one running to it from the axilla and the one running along the clavicle which they call ὠμιαία." (*De venarum arteriarumque dissectione* 2.792.2–4, tr. Goss 1961, 358, where it is identified as the cephalic vein). In humans the cephalic vein runs medial and anterior to the deltoid muscle, not lateral to the deltoid as shown in Vesalius' illustration above.

another branch of the axillary vein forms a vein [*v. mediana cubiti*] similar to it; this vein, made up below the skin from the two branches, we
most correctly call the common vein. Others call it the middle vein.
Although Galen writes that this meeting of veins occurs deep, he had
earlier recorded that it is on the surface, not noticing that the axillary is
divided into two trunks soon after its entry into the arm, and the one
that runs more or less along the skin eventually becomes the one that
Galen took for the whole axillary vein. But the larger trunk, distributing significant offshoots here and there, joined throughout its entry
with the artery to the arm and tucked between two muscles that flex
the forearm, is borne into the forearm, * making the same distribution   76
as the artery (as I have written) as far as the ends of the fingers and not
claiming anything in common with the humeral vein. In fact, you will
find no small number, among which you will discover not even a trace
of the humeral vein around the forearm, in that greater trunk of the
axillary vein that I have described which readily presents branches to
the skin, which Nature uses instead of the humeral.

In addition, the idea must have moved Galen that the large veins
must in every case have been distributed deep in the body to the
radius, the ulna, and the hand, must have been much larger than the
others, and must have preserved more of the same arrangement in all
humans than the superficial veins scattered between the skin and the
fleshy membrane. If this were so, how, I ask, could Nature have been
so negligent that she took the vein [*v. cephalica*] under the skin through
the length of the arm and on its outer side – the humeral vein itself,
from which was then to be made half of the veins which were to be
distributed to the forearm, to the bones of the hand, and in great number to the muscles that surround them? Indeed, we could discuss the
statement of Galen in detail, including the axillary known to him, as
he described it placed entirely beneath the skin above the elbow joint,
from which he then wrote the offshoot originates that deep down
joins with a branch of the humeral vein. So when I regularly demonstrate that we see these things, and do so in a great meeting of the
most learned men with cadavers present and meticulously compare

them to Galen's descriptions, I should certainly have deserved that Nature deprive me of eyes, if I had not protested rather to Galen than to her who is not to be accused of negligence, and I should have showed myself falsely accusing the exquisite designs of Nature and unworthy of my hands and eyes.

Since this dispute about the veins of the arms must be referred rather to faulty descriptions or to things that Galen omitted than to my demonstration that Galen dissected apes and not humans, I shall now abandon it altogether and take my account to other organs.

77 *Reasons taken from the nerves by which it is known that humans were not dissected by Galen*

An account of the distribution of the vessels is too troublesome for me to think that other differences therein should be explained. For this reason, I shall be more brief in explaining differences of the nerves. I shall call attention to only one that should not be neglected: it occurs around the nerves of the sacrum and coccyx. In man, six pairs of nerves belong to the sacrum; the first comes out between the upper part of the sacrum and the lowest lumbar vertebra and has nothing peculiar to it compared to those belonging to the thoracic and lumbar vertebrae. The five lower pairs, however, do not come out in that way from the sacrum, but before they exit the foramen of the sacrum provided for the dorsal medulla, the nerves separate from each other on each side, with one portion going forward and the other to the rear, in the same fashion in which we have said the foramina of the human sacrum are elegantly carved out, with the dorsal medulla coming out of the end of the sacrum in the posterior area. In simians, though, the first pair of nerves of the sacrum, and then the other two, have the same egress as the human nerves, but no more than those three pairs can be ascribed to the sacrum. The dorsal medulla does not end there but passes through the ossicles which Galen identified as the coccyx and which in appearance very much resemble lumbar vertebrae; it sends out the same number of nerve pairs as pass through the ossicles; Galen thought there were three in simians and for that reason ascribed three pairs to the coccyx. But the fact is that no nerve hangs from the actual coccyx in man, nor does the dorsal medulla extend into it.[110] It should not be neglected that in caudate apes and dogs there are not just three pairs belonging to the ossicles of the coccyx, but as many as they have of such ossicles; the nerve pairs must be counted by the increased number.

In Galen's book *De ossibus* the description is applicable to simians in the arrangement of nerves, but not at all to man.[111] In the fragment *On the Dissection of Nerves,* ✳ which I would like not to be ascribed to Galen, or at least to be considered corrupt, no mention is made of those nerves. How well he dissected them when he wrote *De usu partium*, nobody doubts who knows that he then had counted or knew only four bones below the lumbar vertebrae. But because in that work he does not make a great error in counting the number of pairs, I am forced to believe that he there made use of the works of other anatomists who dissected humans. Besides, from the descriptions of the series of nerves going to the arm and the femur as I am accustomed to show it at the universities, I conclude for my part from a comparison with Galen's descriptions that it is very clear that they are not in agreement. Whether that must be in every case blamed on the distinction between simians and humans, I am unwilling to suggest because I have not thoroughly dissected the nerves of simians.

Reasons selected from the contents of the peritoneum

But now it is time to bring into my argument some differences in the viscera and cavities of the body, where the site of the omentum

---

110 For this account of Galen's version of the nerves in the sacrum and coccyx, see *Fabrica* Book IV ch. 16: "Distribution of the Nerves Coming from the Sacrum."

111 Galen wrote "The sacrum is composed of three parts, intrinsic vertebrae, as it were, of its own, at the end of which lies a fourth, another bone, called coccyx. When

first becomes visible (I will pass over in silence the hardness of the peritoneum in pigs and simians, which Galen's dissections thereabouts readily acknowledge in those animals, though because it is thin and soft in humans I am seldom granted my wish to dissect). You will find very few humans indeed in whom the omentum is seen wrapping the intestines as far as the pubic region. But to a great extent it is seen drawn up on its lowest side above the region of the colon, which extends along the stomach for the width of the body; this region descends on the left side more than the right, never reaching the region of the umbilicus, though by using the hands it can be pulled downward to cover the intestines as far as the pubes. But in simians and dogs, it is always so extended over all the intestines that it exists between the bladder and the rectum in males and in females between the uterus and the bladder. I know that humans have been called ἐπιπλοκομισταί[112] by some from the bulk of the omentum and its way of being carried, and that Galen believed for this reason that man has the largest omentum. But I believe in my judgment that pigs, dogs, and simians possess a much greater omentum than man, as I am forced to conjecture from people consumed by wasting disease and again from extremely obese women. An abdomen that stands out a foot and a half in these especially fat humans, as perhaps Galen had

---

they are all cleaned by boiling, the structure [of the sacrum] is seen to be the same as that of the vertebrae. The nerves from the spinal cord, issuing through its foramina, pass out of its would-be "vertebrae" just as they do along the spinal column as a whole, yet not from its sides but internally and externally. There are three pairs of them." (*De ossibus ad tirones* 2.762.6–14, tr. Singer 1952, 772).

[112] "Possessing an omentum." Galen *De optima doctrina* 556 (LSJ). The Greek word is garbled in the Basel edition; the form as given in the Venice 1546 edition is ἐπιπλοκομιστάς (the accusative form). The nominative spelling known to Vesalius was probably ἐπιπλοοκομισταί, as in Galen *De anat. adm.* 2.556.13: "[the omentum] is largest in men and apes. For this reason many men are called 'epiploon carriers.' They give this name to the hernia [*epiplocele*] formed when the omentum breaks into the passage to the testicles." (tr. Singer 1956, 157). The end of *Fabrica* V ch. 4 spells it with a single o.

convinced himself, does not in fact have that mass and enlargement in the least degree from the omentum.[113] Though Galen had that perception, it need not be surprising because it is nowhere clear that he knew that fat is located between the skin and the fleshy membrane, as I mentioned previously.

A more noteworthy difference, from which you may learn that Galen did not dissect a human, comes from the connection of the

Illustration of the omentum from Fabrica V fig. 4. In the upper left, l is the "stem of the portal vein where it issues from the liver, contained and supported by the lower membrane of the omentum." The rim of the omentum, marked by three letters e, is described as the "circle or orifice of the omentum, from which it takes its beginning."

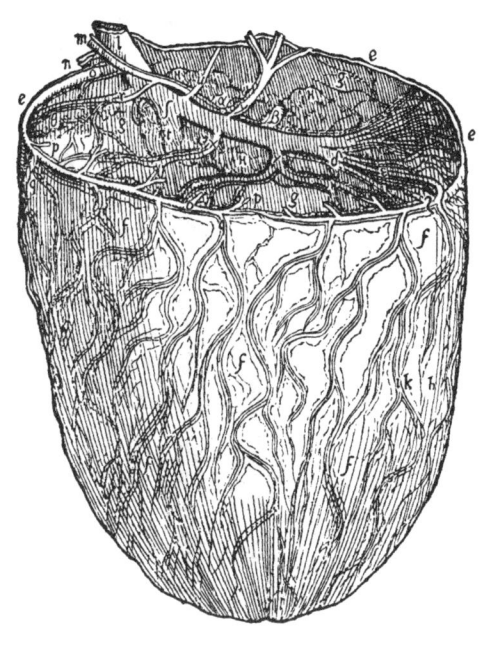

omentum to the colon. In dogs, the omentum is not attached to the intestines by so much as a fiber; but in humans, the entire portion of the colon that extends from the spleen beneath the liver is not attached to the spine by means of any body nor is it contained so to speak by the mesentery; it is supported by the membrane which is not identified as the omentum by Aristotle and many anatomists on the basis of human dissection, but I must take it for the omentum's lower membrane. Just as the lower

---

[113] But see *Gray's Anatomy*, 39th edition (2005) p. 1132: "The greater omentum … always carries some adipose tissue and is a common site for storage of fat in obese individuals, particularly males."

membrane of the omentum in dogs and apes supports the series of the portal vein, arteries, and nerves, the glandular body spread beneath the back of the stomach contains it. This membrane braces the colon to the spine in humans, and then the lower or rather posterior part of

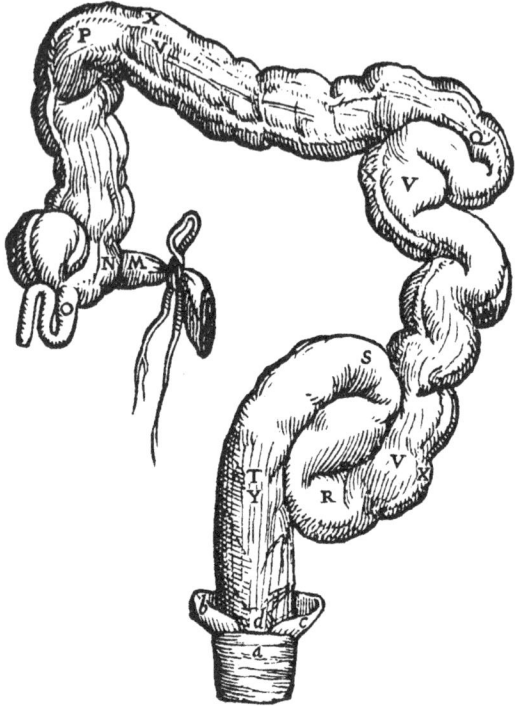

the omentum that covers the intestines beneath the colon takes its beginning, so to speak, from the colon and is not manifestly continuous with that part of the omentum which we said takes the place of the mesentery for the colon here in humans. In simians, however, the continuity appears virtually the same as in dogs, and it is

The colon as illustrated in Fabrica Bk. 5 fig. 8. The cecum is marked O. The 1555 edition of the Fabrica adds "This very thin, small appendix, twisted like a worm, is called by us the blind intestine [intestinum caecum]."

attached to the colon only by a number of fibers, and rarely in simians by a separate connection (as was correctly written by Galen). ⚹

80   Therefore simians have a structure halfway between humans and dogs,[114] and it should be considered perfectly evident that in his

---

[114] Cf. *Fabrica* V ch. 4 on the connection of the omentum with the colon: "The fact that apes have a nature midway between dogs and humans explains why Galen said that the omentum is attached to the colon by a few loose connections, and only on the right side. For in dogs the omentum is nowhere attached to the colon or any other intestine, while in apes it is intermediate, though they are still closer to the construction of the dog than the human."

account of the omentum Galen described the ape and not the human. It will be clearly understood whether this is also the case in the intestines when we have seen that the cecum is spacious and large in dogs and still larger in dormice and squirrels, and in those cases originating as if from the right side of the colon, or continuous with the colon, whereas in man we have seen that the cecum is a thin body twisted like a worm, brought out here from the beginning of the colon more on the left side and positioned, like the cecum of dogs, in the right flank.

If, I say, we have considered these things, we shall no doubt deny that Galen examined the intestines of man, and from his account we agree that all doctors who have followed Galen imagine a certain sac in man, though that thin body in humans should nowhere be counted in the number of large intestines. There should also be no doubt that the ancients called it the blind intestine. But if someone studies the simple series of intestines in a dog near the right kidney and then notices the globular swelling at the beginning of the human colon, the attachment to it of the small intestines, and the beginning of the appendage called the cecum, he will not neglect their need for an extended and elegant description. In my judgment he will immediately admit that Galen had not scrutinized them in man; likewise, he left no account in his books except what he could make to fit a dog. Indeed, if I had not carefully examined the innards of a dog and determined that Galen had written his account only about apes, I would say that Galen had mentioned the cecum because of the opinion of other anatomists. This is because (if I remember correctly) caudate apes differ from dogs in having no cecum, besides the fact that * when Galen enumerates the intestines he makes no mention of the cecum.   81

But if we grant that these are small items, with what negligence, I ask, are we to think that Galen omitted the curvature and course of the colon which it makes toward the umbilicus on the right side where it is conterminous with the rectum, if he had noticed it in man in such a way that he saw that in dogs the colon is carried with no such curvature? He also does not mention the turn that the colon

95

makes above the spleen, which must be fully weighed by doctors when they discuss colic pain on the left side.

In addition, to free myself eventually from the intestines, if Galen had viewed human intestines as much as he did those of the pig and the cow, and had not thought that human intestines resemble those of dogs and apes, we would not read in his writings that the colon of certain animals has powerful ligaments and bulges as if into little spheres on each side. Without doubt he attributed such a form of the colon to man without distinguishing him from animals that have a colon conforming to this image. You should never forget how much the form and structure of the colon in man and the pig differs from the intestine of dogs, which Galen described instead of the human intestine.

We are able to learn from Galen and from anatomists who follow him that the liver is not so conspicuously divided into five lobes or fibres,[115] and embraces the stomach like a hand.[116] It is clear, however, that Galen thought the liver was divided into fibres from the eighteenth [Hippocratic] aphorism of the sixth section;[117] as in simians, dogs, and pigs we see it partitioned in such a way that they do not appear attached by the peculiar substance of the liver. But the human liver is a single, continuous body, in no way divided into fibres. The small cleft by which it admits the vein that leaves the umbilicus is so slight that the liver should by no means be thought divided by it into two fibres ✳, since the cleft does not cut even a sixth of the

82

---

[115] Latin admits two meanings for *fibra*: a filament-like strand, or a part or division, such as a lobe of the liver. The latter carried over into English, though this meaning is now obsolete.

[116] This error of Galen and Vesalius' own predecessors was taken up in *Fabrica* V ch. 7 in the section titled "The form of the human liver differs from that of dogs and pigs." Vesalius himself had illustrated a human liver with five lobes in his *Tabulae Anatomicae* of 1538.

[117] The only Hippocratic text that mentions a five-lobed liver is *De ossium natura* 1.11. This paragraph on the construction of the liver recapitulates *Fabrica* V ch. 7, particularly the section headed "The form of the human liver differs from that of dogs and pigs."

liver's thickness. No one should therefore doubt that Galen had not observed a human liver.

I shall pass over the ligament [*lig. falciforme hepatis*] given to man but not to dogs or caudate apes, which travels in a straight line from the front of the body posteriorly and binds the liver to the transverse septum, crossing from the right side of the vena cava through the septum. Corresponding to this is the attachment of the liver in man to the front of the peritoneum; it is always present in humans, not as in dogs, by means of the vein [*ligamentum teres hepatis*] running from the umbilicus to the liver.

It is the opinion of Galen that this vein and the arteries peculiar to the fetus, together with the passage that carries urine from the bladder between the innermost and the second wrapping of the fetus,

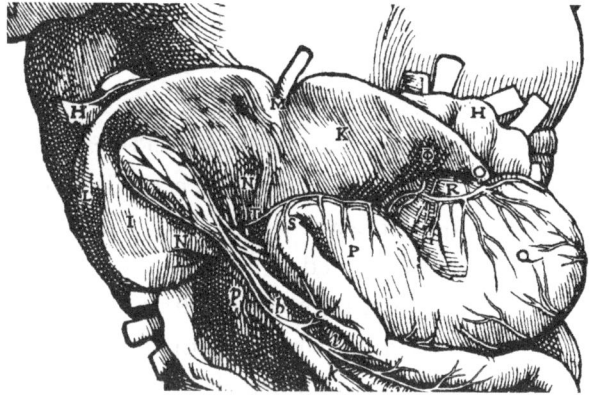

Detail of fig. 12 in Fabrica V, showing the single-lobed liver (K) pulled upward. The obliterated umbilical vein, which after birth becomes the ligamentum teres, is labeled M.

are completely resorbed in animals after birth, just as in dogs that are not very old there is no trace of these vessels. However, in humans into extreme old age those four organs are seen in dissection; they are not in fact hollow and performing the same functions as before birth, but appear like cords and are nurtured by a great deal of fat. Therefore the umbilical vein performs as a very elegant attachment for the human liver, by which it is connected (as I was saying a little earlier) to the front of the peritoneum.

For the same reason, however, the connection of the bladder to the peritoneum is an indicator that the human bladder was not observed by Galen. For we see that the bladder of the dog and the ape, like that of the steer and the sheep, throughout its body or base is not attached to the peritoneum by some process or light membranous ligament, while the entire anterior surface of the human bladder is quite tenaciously attached to the peritoneum and in no place separates from it, except that the aforementioned fetal arteries and the passage provided for fetal urine contribute in no small way to this connection and produce a far different type of attachment to the bladder in humans than in the animals mentioned. *

How important this distinction is to consider I judge to be no trivial matter, because I know from experience that so many wounds that penetrate from the pubes into the space of the bladder, and therefore provide a path for urine, are easily treated. But if the attachment of the bladder to the peritoneum were the same in man as in the dog, and no greater than the connection to the intestines or the stomach, there would be no hope remaining for the treatment of those wounds.

In addition, if Galen had described the site of the spleen and its connection to the stomach a little more completely, we should be able to question whether he believed from the dissection of dogs that the spleen is located so much in the anterior of the body and down by the lowest part of the stomach, and as obscurely attached as most doctors and anatomists believe; for the difference between man and the dog or the ape in this part is not slight.

From the lengthy study of the penis in Galen,[118] a description in which he seems to have taken great satisfaction, and then from

---

[118] Vesalius' disagreement with Galen about the structure of the penis is more clearly stated in his chapter on the penis in *Fabrica* V (ch. 14). In bk. 15 of *De usu partium*, Galen called the corpus cavernosum a "sinewy (νευρῶδες), hollow body growing out devoid of moisture from the bones called pubic… it grows out from bone as all the other ligaments do … and is the only one of them all to be hollow" (4.217.14–218.6, tr. May 1968, 658). The "hollow" in Galen's description probably refers to the corpora cavernosa which flank the corpus spongiosum and the penile urethra (but do not in fact grow out of the pubic bones). Vesalius himself believed that the

a painstaking dissection of the male member, nothing is more clear than that it had never been examined by Galen. While carefully investigating the method of its composition, and bearing in mind that of cattle and bulls, which we employ in preparing straps and sometimes certain shackles, I learn that it fashions some hollow sinew, and brings forth many other things having little resemblance to the two bodies that principally form the human penis; these consist of a spongy and somewhat fleshy substance enfolded in a very thick membrane. Aristotle,[119] or the person from whose account he wrote anatomy, appears not to have noticed this even in passing. So nothing will more manifestly show that Galen's description is at variance with the construction of man in many respects than my own quite lengthy account, which cannot be repeated here because of its length, or a comparison in dissection with the human penis; nothing is more manifest than the things I have plainly explained, which are evident without dissection – those matters came to be scrutinized because of him. ✳

As the uterus has a great number of features worth our attention in its position, shape, attachment, substance, insertion of vessels, and in all the things that we examine in the study of the parts, it has a great many features worthy of attention, which cannot be contained in an easy or short description. They come primarily from Galen, in Attic Greek, and they did not escape hard work in writing. It should not seem strange if when he wrote the special book *De uteri dissectione*, many things come to mind from my general conjecture that he did not dissect humans, that add strength to my opinion.

This first thing to be noticed in his book *De uteri dissectione* testifies more than once that the uterus of the cow and the goat is,

---

corpus cavernosum originates from the pubic bone; Galen's error therefore lay in implying that there is only a single such body. The *ligamentum suspensorium penis* is the only part originating from the pubic bone, but it is unmentioned by Galen or Vesalius.

[119] In bk. 1 ch. 13 of *Historia animalium* (493a26), the fleshy part Aristotle mentions is only the tip or glans. The rest of the penis, he says, is cartilaginous, χονδρῶδες. This describes the penis of certain animals but not of man.

as predetermined, like that of a woman, and he does not mention any difference between those uteri, except that because of the authority of other experts in dissection he dared not contradict anyone openly for what they had written privately to anyone about the human uterus that did not occur also in their bovine uteri. Galen says that they should not lightly be contradicted because they acquired their knowledge of anatomy not in brute animals but in humans.[120] He does not make the same claim for himself but brings his

Illustration of the Galenic bovine uterus from Fabrica V fig. 29, showing the division into two horns that most distinguishes it from the human uterus. "In the end," wrote Vesalius, "the student will learn that Galen never inspected a woman's uterus even in his dreams; he saw only those of cows, goats, and sheep." (Bk. V ch. 15).

apes into his account instead of women. Though some professors of anatomy mentioned by Galen had described the uterus of goats and cows (Praxagoras was one of them), passing on contradictions of earlier writers, those present at my dissections would observe in the medical schools with some pleasure how anxious Galen appeared about those whose opinions he must agree with and what had to be universally

---

120 See Galen *De uteri dissectione* 2.895.7–14: "'Not in the case of all women but in some,' says Herophilus, 'four other vessels branch off from those that go to the kidneys, and enter the uterus.' This I did not find in other living beings, except occasionally in apes. I do not, however, disbelieve the fact that Herophilus often found them in women. For he was not only competent in other branches of the art [of

established about the uterus. Because he found the descriptions by later writers in cows, he attributed these only to earlier writers and knew the descriptions were not of dissected women, but he was not willing to dispute their opinion openly.

Since, therefore, the human uterus differs greatly from the bovine, proofs are easy to find ✳ by which it can be clearly known that Galen did not dissect a human uterus. The cow's uterus goes beyond the position of the sacrum, angling downward from the rectum and above the vessels running upward along the loins where the lowest lumbar vertebrae stand. Thus Galen's description squares here with the human, though the non-pregnant human uterus does not ascend so high in position that it comes within a long distance of reaching the point where the sacrum is joined to the lowest lumbar vertebra. Nor does it go upward beyond the bladder, as do the uteri of cows and dogs. The cervix or neck [*vagina*] of the uterus in quadrupeds is extremely long, and straighter than that of the human uterus. It will therefore be surprising to no one that for this reason its position is different. From this different position, we cannot be in doubt that Galen measured the length of the neck of the woman's uterus only in fingers from the length of the man's penis. Otherwise

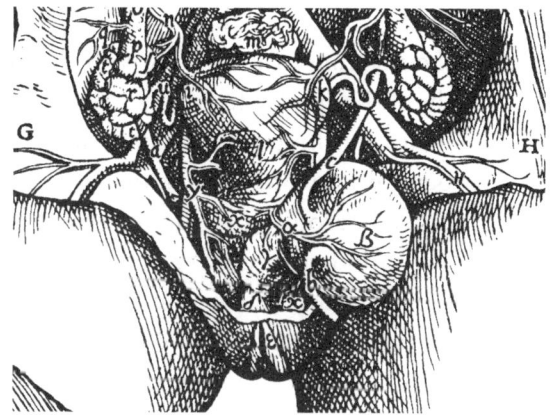

Detail of fig. 25 in Fabrica V, showing the human uterus (i, k, l) "somewhat round, like the bladder" (α, β), which has been turned forward to the left.

medicine], but he attained the highest degree of accuracy in things which become known by dissection and he obtained the greater part of his new knowledge, not, like the majority [of physicians], from irrational animals but from human beings themselves." (tr. von Staden 1989, 220).

he would not have ascribed such a high position to the uterus, if the neck stood together with the uterus. In describing the form of the uterus, Galen seems to agree with the opinion of Herophilus when he compares it with the bladder. But later he no longer agrees with himself when in arguing about the horns he to some extent makes his own use of Herophilus' words that fit women. As he proceeds, he judges by a different opinion expressed by anatomists, and then as he affirms that the bovine uterus is like the human he disgraces himself by wrongly describing the horns and shape of the human uterus differently from Herophilus.

The human uterus is somewhat round, like the bladder, compressed in front and behind and accordingly wider than it is deep. On its upper side, it also appears somewhat depressed, and resembles a quarter-circle or crescent moon. Otherwise, if it bulged like the bladder, it would form a semicircle of a complete circle. So the upper part of the uterus ends in a kind of swellings like those on the head of a calf because it will soon bring forth horns. These swellings or peaks were probably what Herophilus called the horns ✳ on the woman's uterus. In dogs and pigs the uterus is split into two parts immediately after the meeting of its fundus with the cervix [*vagina*], as if you separated the index finger from the middle finger as far as possible and imagined each finger to be one part of the uterus. These do not appear bent in non-pregnant animals, but in those that are pregnant they are curved according to the number of fetuses. Those are contained one after another in these parts of the uterus; hence those who have told the anatomy of such animals from dissection have called those parts of the uterus horns because they closely resemble a large and prominent horn, and they say fetuses are conveniently contained and enclosed by them.

In cows, however, everything is different. If you look at the fundus of the uterus in one piece and with its outermost wrapping still in place, it is seen to be simple, borne from its neck upwards on a long course, not round in shape but rather thinly rounded in front and behind and compressed as if two fingers were placed against each other and covered by a fascia. Then the fundus is separated as if into two ram's horns separating from each other. I do not know whether you would liken anything in the entire body more similar to something else than you could compare these parts of the cow's uterus to a ram's horns.

And since the fetus is located lower down in the other part of the fundus (for it is double, as I shall soon say) lower than in the separation into the uterine horns, and since therefore the fetus is so to speak not contained in them, why is it surprising that Galen is so muddled in his description of the horns? Why is it surprising that there are conflicting doctrines of anatomists about the quite different kinds of uterus? On the other hand, it is a fine and intelligent thing that he thought himself extricated when he believed that the action and function of the horns should be assigned to another place which you will never find in his writings.

The entry of seminal vessels occurs for all uteri in its farthest peaks; in women, it occurs in the protuberant heads on each side of the upper part of the uterus;[121] in dogs, at protuberances of the top of the parts which we have compared to fingers separated from each other. ✳

As in cows, the oviducts are beautifully folded into the peaks of the parts of the uterus that resemble ram's horns. As Galen has also well noticed, this feature was missed by other anatomists.

87

I cannot sufficiently judge, however, what vessels Galen writes about that are inserted into the neck of the bladder and were known to Herophilus and others whom he was afraid to contradict because of the great precision that marks their description of other parts.[122] It appears Galen believed that in women as in men what we call the

---

[121] At i and k in the preceding illustration.

[122] A reference to the pampiniform venous plexus in the male. In *Fabrica* V ch. 13, he writes "if I rightly understand Galen's view in *De semine*, in one place he wants these turnings to be called the κιρσοειδής παραστάτης as Herophilus believed,

glandular assistant was attached to the neck of the female bladder, and then discovered the vessels that are implanted from the ovaries[123] into the neck of the bladder at that body in women. If Galen read that in Herophilus, and he himself held the same view, I am willing to say that I found that glandular body but I did not also find the seminal vessels brought from the ovaries in women or in animals. Indeed, I shall distrust the authority of Galen in this part until I have heard that such things have been found by someone in bodies and am told by what means I may study those matters. Meanwhile, I consider none of them a necessity for women.

I would not want to fabricate anything here about the coitus of pregnant women, about whom it is universally agreed that the mouth of the uterus is tightly closed. My opinion is the same regarding the veins and arteries going to the uterus, except for the vessels delivering the material of semen from those that are borne into the kidneys. Herophilus said that these are seen in some women, so Galen did not doubt that he had found them in many women.[124] He would have written otherwise if he had ever dissected a woman himself, and had observed that they are absent exactly as they are in apes. When

---

while elsewhere he applies that name to another part of the body we are now describing, where it is inserted in the penis." In *De semine* 4.565.14, Galen identified this term with the ductus deferens as a whole; in 4.567.13, he applied it to the distended end or ampulla of the ductus deferens (or perhaps the seminal vesicle) at the base of the bladder. In 4.582.15 he says "Herophilus gave the name 'varicose helper' to the part of this vessel that lies close to the penis." In 4.587.1 ff. he again identifies it with the ampulla: "In animals that have abstained from congress with the female all the parts are full of semen, first the varicose helper, then the entire spermatic duct, then the epididymis, then the entire testicle, etc." (tr. De Lacy 1992, 117, 121, 135, 139).

[123] Lat. *a testibus*; for the sake of clarity we are translating Vesalius' female "testes" as the ovaries. It should likewise be remembered that the female "seminal vessels" are the oviducts.

[124] In *Fabrica* V ch. 15, Vesalius had written "In many places Galen agrees with Herophilus (who he knew had dissected human bodies) almost against his own judgment, not willingly admitting – though he had never seen one – that he was not describing a human uterus." He refers there to Galen *De uteri dissectione* 2.895.7–14 (see n. 120 above).

he says that he had found those veins and arteries only on the rarest occasions in simians, I am convinced he had written that in the same way I do when describing a different set of nerves and vessels in some places and am compelled to add that it occurs sometimes in one way and sometimes another, * because I am at first afraid to contradict   88
Galen even though I had always observed otherwise. Similar to that are places in my writings regarding the origin of the humeral vein from the vena cava, the distribution of nerves in the lower leg and the foot, and the meeting of certain nerves in the upper arm.

However, I must return to the subject of the uterus, whose contact with other parts proves beyond doubt that Galen did not see its position and shape and only imagined how they are in women. He describes its size inaccurately, thinking that the fundus of the uterus in women is much larger than it appears in reality; I never saw the uterus of a non-pregnant woman to which I could assign a breadth of three fingers, nor did the length of the fundus ever exceed three fingers' width. The length of the vagina proved itself variable proportionate to its extension, for it is a membranous, sinewy body and accordingly more pliant than some would suppose. I therefore believe, from the difference between the woman's pudendum and that of cows and goats, that Galen thought the human vagina was equally muscular and fleshy, and described it differently than it is in fact.

Again, as Galen correctly described two cavities in the bovine uterus whereas it is a single one in a woman, so too he rightly described the tunics of the bovine uterus but was quite wrong in the construction of the human uterus. There is a single tunic in the woman, which is wrapped differently from the peritoneum and the ligaments containing the uterus. In cows, however, as we said the fundus of the uterus is immediately divided into two parts, so too each part is formed with its own peculiar tunic. This is why Galen put it out that the uterine tunic is in two parts, one of which is covered by the peritoneum, and it has this instead of only one because it wraps the two parts of the fundus at one time so that no seam will come between the two parts like a fascia with which you would wrap two fingers or the roots and beginnings of a ram's horns together. As the uterus * is shaped, so too is the difference in the tunics because of the division of the uterus, where Galen did not know that the human uterus differs from the bovine.

For the present, while wishing to avoid lengthening my remarks about the uterus too far, I must not pass over the place in his book *On the Dissection of the Uterus* where Galen seems to me as worried and perplexed as we sometimes feel when our mind changes and is driven one way and another whenever we are overwhelmed by the opposing opinions of authors. Galen had read that the leading professors of dissection held that the human uterus had

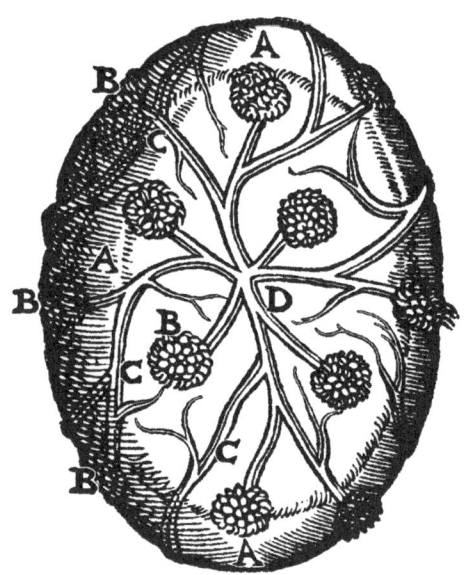

Vesalius added this figure of the bovine fetus and its wrappings to the 1555 edition of the Fabrica to illustrate the cotyledons or acetabula of ruminants (marked B): "We add this figure to those supplied to demonstrate the parts in man, to show the difference between the fetal wrappings of cows, bovines, goats, deer, and other horned animals, as well as animals truly showing acetabula in their uterus after conception."

no acetabula,[125] but attributed them to the uteri of horned animals; others, however, who thought as Galen did that the bovine uterus was like the human, held that the uterus did have acetabula. Hippocrates

---

[125] Cotyledons [*tunica mucosa, glandulae uterinae*]. The discussion of acetabula that follows is related to Bk. V ch. 16 of the *Fabrica*, "On the Acetabula of the Uterus." Such acetabula, pea-shaped pits in the cavity of bovine and canine uteri, are to be distinguished from deep cavities in bones, such as the cup-shaped socket in the hip bone that receives the heard of the femur.

agreed with this: in his *Aphorisms* he mentions acetabula,[126] meaning something much different by that name than whoever established this name in dissections, as no one need be in doubt who has undertaken the dissection of various uteri.[127] It is quite clear to me that the name acetabulum was applied by authors to three different things, so that Galen should have recognized from his knowledge of the thing what each one meant. First, we use the name acetabulum in describing round, hemispherical hollow cavities, in this way calling all deep cavities in bones, primarily the cavity of the hip bone in which the head of the femur is received, an acetabulum. The name, as I have used it in my book, is taken from the shape of the herb that we call acetabulum, garden of Venus, or umbilicus, and a measure for vinegar. Now in the uterus of a cow, a ewe, a bovine, or a doe that has recently conceived, numerous pits are seen which protrude with their lips into the cavity of the uterus and are hollowed as if someone had pressed halves of peas into them. Therefore, no one who takes my advice should be in doubt that such depressions were first named acetabula.

Since these arise and are made out of swellings visible in a uterus ✳ that is not pregnant, like the ones that anatomists have compared to veins standing out in the anus from which blood often issues at intervals, it is not surprising that the alternative name acetabulum is applied to those swellings; but that is quite inappropriate, since they are not depressions unless the horned animal has kept its reproductive power for some time.

90

---

[126] The 45th Aphorism: "If moderately well-nourished women miscarry without any obvious cause two or three months after conception, the cotyledons of the womb are full of mucus, and break, being unable to retain the unborn child because of its weight." (Loeb tr. by W. H S. Jones). This Aphorism was cited in Galen's *De uteri dissectione* 2.905.1–906.5 with the following statement by Galen: "The cotyledons are the safe attachment between the placenta and the womb. They say the human womb does not have cotyledons, but they exist in the wombs of cows, goats, deer, and such other animals, being loose bodies, somewhat mucous, in their appearance resembling the cotyledon in grass, the little cymbal, from which they take their name," etc.

[127] The end of this sentence is garbled and cannot be construed with any certainty.

Again, because veins and arteries into the outermost wrapping of the fetus are formed out of those pockets, it should not be a surprise that those veins and arteries newly generated between the wrapping and the inner surface of the uterus are called acetabula by some people. It is all too clear that Hippocrates employed this term when he wrote his account of abortion in the second or third month and explained as the reason (which he otherwise almost always passed over in silence) the mucus with which those new, weak vessels swelled by which the fetus is supported. It is certain that the two-month fetus and to a much smaller degree the three-month fetus even in cows is retained only by those depressions or rough places, just as we all know cuttlefish and squid cling to cliffs and rough places with their proboscis, which we see is full of acetabula.

Because these vessels occur in women as they do in cows, the name acetabulum would for that reason not be wrongly attributed to a woman in that sense. But because the human uterus does not show the swellings standing out in its interior, and because in the absence of swellings the depressions carved out in the shape of the socket in the hip bone cannot exist, it is clear that neither the first nor the second name of the acetabula can be applied to a woman. Galen would therefore not have been confused if he had also occupied himself with a human uterus as opposed to a bovine.

Some conjectures based upon the parts that are contained in the thorax

There is also considerable difference in the wrappings of the fetus, but we shall better save them for the untrue descriptions and direct our attention to the second cavity of the body, the thorax. This will be briefer because of * the aforementioned course of the vena cava, in describing which I stated how the membranes that divide the thorax in dogs and apes constitute a major cavity above the sinewy part of

91

the transverse septum and beneath the wrapping of the canine heart that does not occur in man. Neither is the peculiar lobe or fiber of the lung found there, which occupies that cavity and would support the vena cava like a hand or elegant fulcrum as it passes that way and deliver it safely; Galen called this the fifth fiber of the lung.[128] But in fact you will find two on the right and two on the left side; this cavity and that fiber of the lung clearly prove that Galen dissected apes and dogs but not humans.

The same is true of the wrapping of the heart, which Galen says is separate from the transverse septum and not attached to it; it is clearly as much removed from the septum in dogs and apes as is the afore-mentioned cavity. But in man, the wrapping of the heart is attached and united over a large area to the sinewy region of the transverse septum and occupies an altogether different site in man than in dogs and apes according to the difference in the heart's location. That is because it is necessary for the small arch and wrapping of the heart to have the same location as the septum. It is clearly established by anatomy and reason that the position of the heart is different in humans than in dogs and apes. For if we think about the length of the breastbone in humans and compare that with the canine breastbone, we will readily be aware that the human heart is as a rule proportionately very large and occupies a position not as vertical and along the length of the body as Galen ascribes to the heart in simians and dogs and we see clearly in dissection. Because the base of the human heart occupies

---

[128] The *Fabrica* takes up this error in Galenic anatomy in Bk. VI ch. 7: "I still remember the passage of Galen in the seventh book of *De anatomicis administrationibus* where he says that this fifth lobe of the lung does not escape the notice of those who dissect correctly. He implies that this lobe was unknown to Herophilus and Marinus, as it surely was because they dissected the cadavers of humans, and not, like Galen, those of apes and dogs, in which nothing is easier to see than the present lobe. Likewise, it is perfectly clear to anyone who dissects a human that the fifth lobe is not present." Galen had written "there is also a fifth small lobe in the right lung, a mere offshoot of one of the others." (*De anat. adm.* 2.625.12 ff., tr. Singer 1956, 189).

the middle of the thorax between left and right, and is located some-what transversely, the remaining body of the heart is borne ✳ quite obliquely downward to the left side, leaning on the left side of the transverse septum just as I said the wrapping is attached there to the septum. It is as if the human heart were forced to take this transverse position and were unable to be taken freely straight down.

So we know how the canine heart is placed: it inclines somewhat with its point but not its entire body from an oblique position upward to the left side. Because Galen had observed this position in dogs and apes but had never dissected a human, it should come as no surprise to us that he never mentioned the position of the human heart, and that Galen, not content with a simple explanation of its place, in fact goes to some length to criticize Aristotle and others who revealed its position to us as it is in dissection and as reason also dictates.

Reasons taken from those contained in the skull

So far as the brain is concerned, it is perfectly clear that the brains of cattle were provided by Galen for inspection and dissection, since in his writings there is no mention of another animal's brain. Because there is no great distinction between the brains of cattle, wild animals, other animals, and humans, there are few grounds for conjecture that Galen did not dissect humans in the third cavity of the body, namely the skull. One of these grounds, not the common one, has to do with the position and shape of the cerebellum. In cattle, as Galen's descrip-tion also has it, the cerebellum occupies the back of the occiput, taken to the rear beyond the posterior region of the cerebrum, whereas in man it is placed altogether beneath the cerebrum, and the farthest part of the cerebrum is extended much more to the rear and into the occi-put than the back of the cerebellum.

Moreover, the cerebellum of the steer rises higher in the occi-put than that of man, whose cerebellum is wide and flattened on top, while the bovine cerebellum is round and spherical there and rises to

the higher position⁎ of the suture that resembles a Λ. The top of the   93
human cerebellum, however, stands only in the middle of the distance
that is measured from the foramen carved out in the occipital bone
for the dorsal medulla to the aforementioned location of the suture
resembling a Λ. It is therefore not surprising that in his description
of the place where the first two sinuses of the hard membrane meet,
Galen clearly shows that he is practiced in the dissection of cattle
but not of man. That place of meeting stands at the highest point of
the aforementioned cerebellum, which differs from cattle in humans.
Galen's description fits the former well enough, but does not match
the latter.

It will be told how the network of arteries at the base of the
cerebrum was seen by Galen in cattle more than humans when I
endeavor a little later to show that he refrained from the truth in his
account of things. At present I will seem to have gone too far in listing
the differences between man and the other animals in order to prove
that Galen did not look at the bodies of humans. I shall therefore put
an end to those differences as soon as I have added that the chief reason for surprise in Galen is that in trusting his apes he so often contradicted the Ancient professors of anatomy without good reason and
accused them of negligence on the ground that they either left out or
described differently what he was finding in apes because they were
observing those features in humans.

Some places where Galen openly criticized the Ancients because they had dissected humans and not apes, as he did

A place of this kind presents itself in his description of the sutures of
the upper maxilla, where Galen writes that other anatomists did not
see two sutures, or passed them over as if not worth consideration.
But Galen based this opinion on the ape, while they were correctly
describing the fabric of man. Similar to this is the process in the

111

lumbar vertebrae that apes and dogs have: Galen writes that it had been overlooked and neglected by others, but nothing is more certain than that humans do not have it.

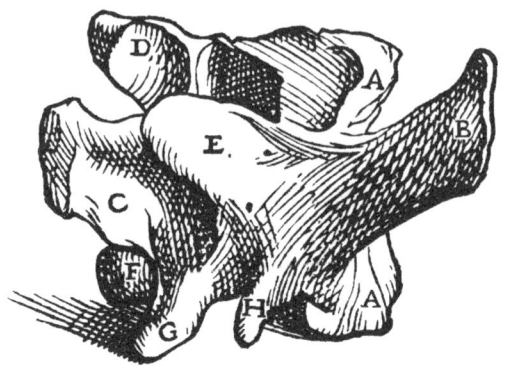

This figure from ch. 17 of Book I of the Fabrica illustrates a lumbar vertebra of the caudate ape, where H marks the small accessory process that Vesalius argued (wrongly) does not occur in humans.

94   They had compared the entire pectoral bone to a sword, but Galen compared the pointed cartilage [*processus xiphoideus*] to a sword because of ✳ his apes and dogs. How frequently and at what length did Galen criticize the ancients for not recognizing the muscle formed from fleshy membrane at the side of the armpit and thorax? But this is not found in humans.[129]

He is no less pleased with himself that they had not observed some of the muscles raising the scapula. But we know from careful dissection that apes have two such muscles.[130] Again, who could

---

[129]   This probably refers to the *m. panniculus carnosus*, about which Vesalius wrote "Galen ... boasts continually that all the authorities in anatomy who preceded him had missed this muscle (while working on human bodies rather than on apes, like Galen), and that he had found it in apes. I have definitely seen it in apes, dogs, and many quadrupeds, but it is not equally wide or fleshy in all of them." See *Fabrica* II ch. 23. By structuring his polemic in this Epistle as a defense of the ancients against the strictures of Galen, Vesalius assumes the antiquarian mantle of the Humanists, whose faith in *prisca medicina* was unquestioned.

[130]   See the section titled "The muscle raising the scapula that is found in the ape" in *Fabrica* II ch. 26: "if you believe Galen you will blame yourself for ignorance and carelessness while dissecting when you go in search of additional muscles in man that lift the scapula."

excuse Galen for ascribing some negligence to others regarding the second pair of muscles moving the head? This muscle is positioned so differently in man from the ape or the dog, and Galen describes it no less than the muscles of man.[131] It is also known about the muscle inserted from the clavicle and pectoral bone into the mammillary process of the head and serving motions of the head, in describing which Galen attacks the ancients quite undeservedly on account of his apes.

So too it is not surprising that Galen found fault with others for passing over the muscle that we see is part of the rectus abdominis muscle in caudate monkeys and dogs, running from the costal cartilages along the side of the pectoral bone to the clavicle, which humans lack altogether.[132]

For the same reason he criticizes other anatomists because they did not know of the muscle presented to the ribs from the upper cervical vertebrae, descending along the anterior surface of the muscle [*m. serratus anterior*] identified as the second of those moving the thorax; it is quite evident that this is lacking in man.

When he mentions a portion of the larger muscle [*m. adductor magnus*] which I count the fifth of those that move the femur and not among the movers of the tibia, and which is inserted by a tendon in the inner head of the femur, he criticizes others because they wrote

---

[131] See *Fabrica* II ch. 28; in the section titled "The second [pair] is quite various," Vesalius remarked "Galen is without doubt seen to testify that many professors of anatomy counted not just two muscles of this pair: some numbered four, others six in the same area, though these were perhaps from the ranks of those who unlike Galen studied humans. In apes and dogs (about which I have spoken so far in describing this second pair) I would count only one muscle on each side, since if one had to pay attention to the impressions one would have to count the same number of muscles as there are origins from the transverse processes of the vertebrae and insertions into their spines. In man, I have observed that this combination of muscles is much different than Galen described it, or rather counted it."

[132] See *Fabrica* II ch. 24, where in the section titled "Dissection of the first muscle moving the arm" Vesalius invited students to look for this costal part of the rectus abdominis, which occurs only in apes and dogs.

that this insertion is accomplished by a tendon and not by a fleshy implantation as appeared to him in his apes.

Likewise he carps at others for having written that tendons of muscles moving the tibia inserted in its anterior part are round and not wide as they are seen in simians. In addition, ✳ how often did he attack certain people because they left out the muscle from which the hidden tendon of the foot originates?[133] The fact is that his description does not fit humans, but rather simians. He censures others for not knowing that the fourth muscle moving the foot is implanted in the heel bone with a fleshy insertion, though it has a quite perfect tendon in humans as distinct from simians.

But what more obviously proves the falsity of this accusation by Galen (to make no mention of the position of the heart and other locations of this kind) and his allegation of negligence than the course of the vena cava through the transverse septum and the fiber of the lung that supports it? When Galen cut up his monkeys and saw that they differed from the description of the ancients, who trained themselves on human dissections,[134] he did not scruple to state that they had not seen that fiber of the lung and who knows what else. I should therefore be thought more impious if I had not vindicated those Ancients with a true description of the human fabric. If because of the powerful devotion to Galen under which I labor and my special regard for him I were to leave his opinions everywhere undisturbed contrary to the testimony of my eyes and the truth of the matter, I should be willing to have my generation wander in confusion like all the ages that have followed Galen, and let his misrepresentation of the Greeks go undetected.

---

[133] This is explained in *Fabrica* II ch. 58 "On the Hidden Tendon Attached to the Sole of the Foot."

[134] By the (sometimes capitalized) Ancients, Vesalius tacitly means not the Hippocratics, Aristotle, or Diocles but the pre-Galenic Hellenistic anatomists, of whom he mentions a kind of pleiad consisting of Marinus, Eudemus, Herophilus, Andreas, and Lycus in the Preface of the *Fabrica*.

Not everything in his description of the parts was correctly reported and described by Galen

So it was, my Joachim, that when I responded to Sylvius in a letter such as this which presented itself too abundantly, to the point that he took offense at my argument that humans had not been dissected by Galen, I came to the point that Sylvius asserted that nothing wrong had been said or written by him. Here I first began to show how impious we should be toward Galen if in describing the fabric of man we should impute to him as many false descriptions as we find in his books discrepancies between man and the simian, and we would protect his authority if we preferred to say more than once in explaining those descriptions that he taught Anatomy chiefly from his inspection of simians, ※ and that he erred in determining that simians are too much like humans. I therefore believed that the things that differ in Galen from the human fabric because of his use of simians and brute animals should not be relegated to the category of variant descriptions; but a number of other descriptions, as they presented themselves in the very large field of places little observed by Galen, I placed almost in this category.

96

A number of untrue descriptions in the bones

When Galen based the type of bones on epiphyses, he wrongly attributed epiphyses only to large, massive bones; but digital bones, carpal and tarsal bones, are perfectly well known to have epiphyses. Likewise, every part of the teeth outside the gums is clearly also an epiphysis in children. I say nothing of the vertebrae, which should also be included in the number of smaller bones; certain of these, as pretty well all the thoracic and lumbar vertebrae, we see also to have five epiphyses. These also prove (though Galen thought otherwise) that epiphyses belong not only to the large bones, which contain a large cavity filled with a marrow that is not separated by osseous fibers, for example the humerus, ulna, radius, femur, and the bones of the lower leg, but also to bones in which this cavity is not present. Neither the vertebrae, the scapula, the bones attached to the sides of the sacrum, nor the ribs have this cavity. Yet in those bones epiphyses are distinguished just as they are in the bones that I was saying are marked by that cavity.

With this should be combined nearly everything about which I disagree with Galen in describing differences of bone construction. For example, when Galen composed his account of them in his book *De ossibus*, under synarthrosis or articulation he should not have included suture, harmonia [*sutura plana*], and gomphosis, since they should have been placed under a single category together with symphysis because it contains bone structures fabricated by Nature for no motion whatever. Opposed to that is the category of structure built for motion that is either manifest or obscure.[135] *

97   Synarthrosis differs from diarthrosis only in the obscurity of motion, and both join in the same types of structure, namely enarthrosis, arthrodia, and ginglymus. So too it is of no relevance to symphysis that bones either are joined by means of another body or by none.[136] For while he identified three categories of such bodies by whose aid or intervention bones are built together—cartilage, muscle, and ligament—calling three species of juncture by their own names, we can give the name of symphysis to no joint which we would truly say comes together by means of a ligament or muscle.

---

[135]   Vesalius' taxonomy of joints is presented by a diagram in *Fabrica* I ch. 4.

[136]   Cf. *Fabrica* I ch. 4: "I have also departed further from Galen's opinion in saying that all attachments of bones are either aided by some substance, or by none; Galen attributed this only to symphysis in *De ossibus*, and he counted synneurosis, syssarcosis, and synchondrosis only as types of symphysis. His own account gave me the first reason for not following him, where he taught that soft, spongy bones are attached to each other by symphysis with nothing in between, but drier and denser bones come together by means of intermediate materials."

It is no surprise that while this explanation of attachment applies to all bone structures in which we easily see sutures and harmonies and which in advanced age form symphyses by means only of structure and attachment and are attached without the aid of a third body, actual joints are held together by ligaments laid upon the bones, going around them, and often becoming involved by actual attachment; they are then held in place by muscles (which are considered as flesh). In younger people we see cartilage quite handsomely come together into a symphysis.

When it comes to mind that Galen in talking about symphysis wrote that spongy, soft bones are attached with no intervening substance while drier, denser bones coalesce with something intermediate, we nevertheless see, as I recently said, that in children, whose bones are spongy and soft, symphyses are formed through the medium of cartilage while in the elderly, whose bones have become arid, dry, and more dense, such cartilage has altogether disappeared; indeed, not even a line of attachment generally appears.

As Galen had little thought of types of bone attachments or constructions, so too he appears skimpy in providing examples with which to confirm his joints. Yet when examples are cited of the types of joints known to him, errors are not altogether missing. *

He shows himself to be not much of an anatomist when he   98
groups vertebrae with ginglymus and thoughtlessly states that the highest and the middle vertebrae are attached by ginglymus because they both support and are supported, not considering that this distinction exists in various joints and that he cites three bones to explain the composition of ginglymus. For although all the vertebrae that there are up to the first cervical vertebra above the twelfth thoracic (or whichever one there is received on both sides) lie beneath a vertebra on their upper surface, they also admit one that is placed under their lower surface. Those in turn that are in the loins are articulated in the opposite way; but in the articulation of two vertebrae to each other, a mutual entry does not arise, and for that reason neither is

there a ginglymus. Though one might wish therefore for a bone to be attached by ginglymus because it admits a bone on its upper side but another enters on its lower side, or conversely it receives on its lower side and is inserted on its upper, a person who has no system for the composition of a joint and does not observe how many bones he is bringing to one example or one construction will pile up no small number of bones, such as the first digital bone, as being joined by ginglymus, and like Galen will make quite conflicting statements, since the types of joint must always be applied to the attachment of two bones.

But we shall set these aside and move on to specific descriptions of bones. The bone of the head numbered eighth [os ethmoidale] comes to mind, placed at the base of the frontal bone, which I am surprised Galen did not surround with his peculiar sutures. I wonder why he thought it pervious more like a sponge than a sieve, since its foramina are more comparable to a sieve, and they were not provided specifically for the dripping of cerebral phlegm, since phlegm could not be taken this way without disease. But this opinion of Galen should better be considered in the bone resembling a wedge [os sphenoidale], which presents the appearance given to a sponge and is constructed below the base of the cerebrum with various foramina that do not penetrate directly, ⁎ like those of a sponge or rough pumice. Galen's view, however, is false, and the entire surface visible in the space of the skull appears smooth (except for the sinus and its processes), and is formed whole no less than the bone of the forehead, the temples, or the occiput, not of a crust pierced in the manner of a sponge.

It is definitely a surprise that Galen imagined such a construction to be for the purpose of straining phlegm, which would have been a very poor design, and that he should not instead have discovered the hollows which occur with great frequency in that bone at the place where it needed to be thick, among other reasons for the sinus where the gland [hypophysis (glandula pituitaria)] is contained that receives cerebral phlegm and is everywhere joined to such elegant processes that are no less important to observe than the sinus itself.

Indeed, from the variable and noteworthy composition that occurs in the cavity of the skull, namely the many depressions and protrusions made for the greatest usefulness by the supreme Creator of things with signal foresight, I conclude that Galen considered nothing because he mentioned none of them and wrote nothing of the things appearing there; perhaps someone would argue that he mentioned the bone that is observed located in the cerebrum.[137] It appears that Galen found some mention of such bone among the professors of dissection, and to avoid seeming to have overlooked or omitted anything said by others, mentioned that also very much in passing.

For my part I find that the human cerebrum has no bone; I first learned of one in dogs, from which talk of this bone made its way to anatomists. In dogs, a large, wide, but thin bone [*tentorium cerebelli osseum*] stands between the cerebrum and the cerebellum along the lower hollow swelling of the cerebellum, convex superiorly for the cerebral concavity and brought forth from the occipital bone to which it is completely continuous. It can be deduced that anatomists attributed this bone to certain animals, and that Galen, though it was not clear to him what it was, ✳ mentioned it.

100

It is the same with the bone that Galen imagined from Hippocrates is in addition to the acromion and the clavicle and is the third in that joint.[138]

There is no account in Galen of the cavity [*cavitas tympanica*] carved out in the temporal bone for the organ of hearing. But it is so complex and elegant and attests the craftsmanship of the supreme Maker to the highest degree, if anything does.

---

[137]   See "A bone inside the canine skull" in *Fabrica* I ch. 6.
[138]   In *De ossibus* 766.9–12 Galen described the diarthrosis of the acromion and clavicle: "Some anatomists call this combination *acromion*, but others maintain that besides these two conjoined bones there is a third bone, found only in man, which they also call *katakleis* or *acromion*" (tr. Singer 1952, §775).

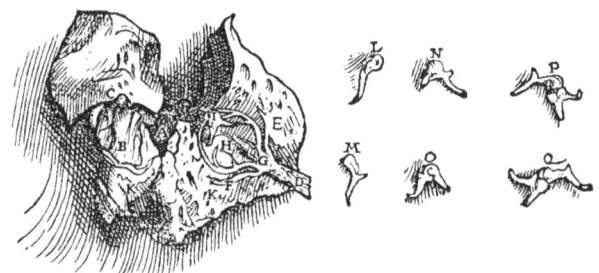

The tympanic cavity in the temporal bone as illustrated in Fabrica I ch. 8 with the two auditory ossicles that Vesalius knew: the malleus (L, M) and the incus (I, N, O). In P, Q they are shown as joined in the ear.

As that cavity escaped his notice, so too it is learned that he had no knowledge of its membranes and of the ossicles, one of which we compare to a femoral bone without its lower heads or to a hammer; we have said the other is like an anvil or a molar with two roots.

In enumerating the roots of teeth, Galen is not entirely accurate: the innermost and last molars have shorter, more compact, and often fewer roots than those that come next to them in order. So too the molars next to the canine teeth do not have as many roots as Galen attributed to them. As he gave slight attention to the roots, so too he neglected the cavity that is handsomely visible in each of the teeth.[139] It is not so surprising, however, that Galen passed over the dental cavity that must otherwise receive much attention when they decay, for he went so far in denying there are hollows in the small bones as to say that the digital bones are solid, though that is not consistent with the truth. They have the same structure as the femur or the humerus in the hollow extended along their length, the receptacle of the marrow.

In his account of the foramina of the head, among other things he wrote inaccurately that the foramen providing a path for the optic nerve is twice as large as the one that transmits the nerve of the second pair.[140]

---

[139] These criticisms of Galen are not present in the *Fabrica*, where much of what is said in Bk. I ch. 11 regarding the shape and arrangement of teeth is paraphrased from Book 11 of Galen's *De usu partium*, beginning at 3.868.19 (May 1968, 516ff.).

[140] *N. oculomotorius*, in modern anatomy the third cranial nerve.

In writing that, he was no doubt thinking only of the nerves. If he had undertaken a careful dissection of this place in man, he would have learned that not only the nerve of the second pair but also a portion of the third, which we know is directed to the eye socket, ✳ is 101 borne through that foramen. Moreover, he would have learned that a large portion of cerebral phlegm is taken this way, and that it provides a way for the vein extending from the eye socket into the hard cerebral membrane as well as for the large artery leaving the skull cavity, and therefore that this foramen is justly seen to be larger than the one by which the optic nerve is taken.

Just as Galen did not observe the course of phlegm, so too on the basis of his description of the reticular plexus[141] we have no doubt that he believed no arterial offshoot is borne to the eyes from the skull cavity; since something of the carotid artery enters the head, Galen contends it is taken into this plexus, from which in turn two branches emerge as large as the carotid artery was when it entered the skull. It will be explained later how absurd these statements are. Now I must confine my account for a time to the subject of bones.

Galen mentions few foramina from a very large number, and he describes them with little accuracy. Such is the foramen by which he says the carotid artery is taken with the third pair of cerebral nerves. The fact is that the artery has its own elegant foramen that deserves the greatest attention. How the foramina of the remaining arteries entering the skull escaped his notice can be known from the arteries of which he is seen to have known none except the carotid artery presented to his reticular plexus.

In his description of the upper maxilla, one cannot agree with Galen. He imagines it is solid, lacking in marrow, and light in weight

---

[141] On the mythic Galenic *rete mirabile* or retiform plexus, see *Fabrica* III ch. 14, "What is Galen's reticular plexus?" Even Jacob Sylvius (*In Hippocratis et Galeni physiologiae anatomicam isagoge*, written about the same time as Vesalius' *Fabrica*) acknowledged its absence in humans but implied that human anatomy had changed since Galen: it "still appears today in brutes" (fol. 57r).

because it is not moved like the lower maxilla; for besides the foramina of the nostrils, which are so large and spreading, large cavities mostly filled with air are seen, and for this reason the maxilla is very light, in more or less the same way that we know images are cast in hollow molds and hung in churches to the saints: when eyes are hung upon St. Lucy, the breasts and chest upon St. Agatha, and arms and legs upon St. Anthony.[142] ✳

102   So if one were to think about the constitution of the mouth, the series of teeth in the upper maxilla, and the seat of the eyes, and therefore know this bone is not small, he will certainly praise Nature, because she did not burden man unnecessarily but at the same time wisely provided for its strength and light weight (which otherwise would have consisted in bulk).

Mention was made earlier of sutures because those that are seen in simians and dogs are not seen in humans. However, it was also well enough indicated at the time that their course was not quite rightly described by Galen even in simians. In certain other sutures the course is so unclear that we learn he described one and the same suture as seen in different places with the name of three sutures.

I do not understand why Galen said that the hyoid bone is not constructed of many different ossicles.[143]

---

[142] In *Fabrica* I ch. 9 Vesalius says "The [upper maxilla] makes up most of the sides and lower area of the nostrils, and near these sides it is quite hollow and not at all solid, but rather elegantly like waxen images that are hollow inside" – perhaps a muddled reference to figurines crafted by the lost wax technique. Saint Lucy is the patron saint of the blind; Saint Agatha is (*inter alia*) the patron of wet nurses and was often depicted carrying her breasts on a plate; Saint Anthony of Padua was the patron of lost and stolen things.

[143] Though the human hyoid bone is composed of five ossicles before they fuse in later life, Vesalius' illustrations of the hyoid bone in *Fabrica* I ch. 13 are more typically animal than human. Vesalius is described as using "the larynx of an ox and of some other animals" in a 1540 anatomy lecture at Bologna "because, he said, in the hanged [human] subjects we cannot see the larynxes, for they are destroyed by the noose, but they are however quite different [in man and in animals]." (Eriksson 1959, 285).

In the same way, in the joining of the thoracic vertebrae at their bodies, I see that the body of no vertebra is formed on its lower side like the cervical vertebrae or received by a concave upper side of the vertebra below it, though that was the view of Galen.[144] The thoracic vertebrae are joined with straight, level surfaces in the structure of their bodies no less than the lumbar.

In a not altogether different way, he stated that all the sinuses of the transverse processes of the thoracic vertebrae by which the second attachment of the ribs is accomplished face downward. But the processes of the lower vertebrae to which the ribs are joined in their second attachment have their sinus facing handsomely upward no less than the sinus of the higher vertebrae inclines downward. Similarly the middle vertebrae have the sinuses in their transverse processes inclining neither up nor down.

It is read more than once in Galen that all the ribs are joined to the vertebrae by two attachments: one to their body and the other to their transverse process, but that the tenth vertebra and the two lying immediately below it ＊ are without transverse processes.[145] The tenth rib is attached by two connections, so it is clear that the tenth vertebra does not lack transverse processes. However, the two ribs below this one are quite small and travel transversely and very slightly forward at an angle; for these reasons it was not necessary that transverse processes be provided for the vertebrae to which they are attached. Nevertheless, for the sake of muscles those vertebrae do have transverse processes, but they are wide and project very little.

I have little idea what came into Galen's mind when he thought that the humerus is the largest of all bones after the femur.[146] In us, the

---

[144] In *Fabrica* I ch. 16 Vesalius had faulted Galen for failing to distinguish the articulations of thoracic vertebrae from those of the cervical.

[145] Vesalius misrepresented Galen here and in *Fabrica* I ch. 16; Galen says three times in *De usu partium* 4. 78–83 that only the last two thoracic vertebrae (the 11th and 12th) lack transverse processes. This criticism was deleted from the 1555 edition of the *Fabrica*.

[146] Galen *De usu partium* 3.121.4 (tr. May 1968, 131), where the humerus and the femur are called "the largest of the members." See also *De ossibus* 767, where Galen

fibula is longer than the humerus, and the tibia is seen to be not only longer but also thicker, at least in humans; in dogs it is somewhat less thick than the humerus. Moreover, the bones attached to the sides of the sacrum are much heavier than the humerus.

When Galen writes about the lower end of the humerus he makes its inner tubercle [*epicondylus medialis*] greater than its outer, calling it inner because it stands at the inner side of the joint of the ulna with the humerus and provides a beginning for many muscles and handsomely serves certain descending nerves. He names the outer tubercle [*condylus humeri, capitulum humeri*] the one that is placed at the outer side of the aforementioned joint when it articulates with the radius and likewise provides origins to muscles; it is many times greater in both thickness and power than the inner tubercle, however much Galen declares to the contrary.[147]

Galen's description of the wrist's articulation with the bones of the lower arm should be considered in detail by the student of the truth; the joint is contrived by Nature's considerable craft, but Galen does not understand its construction. Indeed, he mixes in many things that argue the negligence of nature. Though he believed the entire articulation of the wrist to the radius should be reported as well as possible because the wrist should be moved to a prone and supine position with the entire hand in secondary motion by means of the radius, ✳ yet the ulna upon which the radius is then moved to a great extent supports the same motion; Nature has had the foresight to augment the radius beyond its usual thickness near the wrist and carved a long depression into it where the three upper bones of the wrist, adhering to each other and constructed as a single head, are

104

---

calls the humerus "the largest bone except the femur." (tr. Singer 1952, 773). Vesalius corrected this in *Fabrica* I ch. 23.

[147] In *Fabrica* I ch. 23, Vesalius cited Galen *De ossibus* chapter 16. See 767–8: "The lower end [of the humerus] divides into unequal condyles. The radius diarthroses with the outer, but no bone engages the inner, so that it seems much larger than the outer, though it is but little larger." (tr. Singer 1952, 773).

articulated. This is also done while a portion of the third bone extends beyond the radius and rests in the ulna (though not without the aid of another body).

For because the radius was unable to constitute the entire socket, it extended a cartilage on its lower

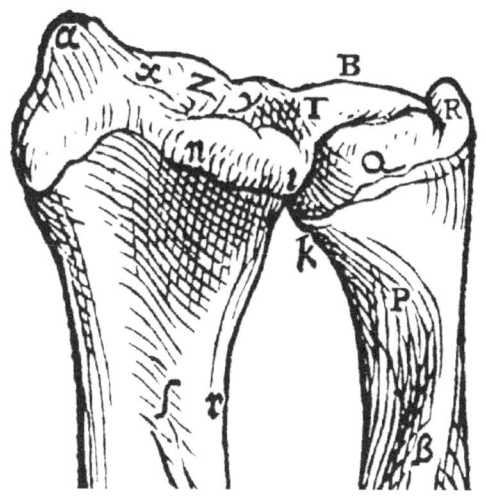

Detail of Vesalius' first figure in Fabrica I ch. 24 showing the distal end of the radius (on left) and the ulna (right) where they articulate with the carpal bones. The articular depression or socket is marked x, y. T is the articular disc, the cartilaginous extension of the radius as far as the acute (styloid) process (R) of the ulna.

side which is smooth and slippery on its lower surface and rests upon the round head of the ulna as far as the inner side of its acute process, taking up that head of the ulna like the socket of a joint. This cartilage [*discus articularis*] originating from the radius below is smooth and slippery, fitting and admitting the rest of the third carpal bone [*os triquetrum*] and filling in whatever was lacking from the radius in the depression where it was articulated to the wrist, since the wrist is not contiguous with the ulna to form a joint without the assistance of another body. A ligament comes forth from the tip of the acute process of the ulna, which is not covered by the aforesaid cartilage, in the same way as ligaments originate from the highest peak of the radius and the surrounding edges of the socket, attaching the wrist to the bones of the forearm. This tip of the ulna is not smooth, and makes contact with the wrist as a kind of joint. But it seemed otherwise to Galen, who was overly partial to the acute process of the ulna, and described the articulation of the wrist to the forearm (so far as I understand from his description) as equally belonging to the radius and ulna. How far that is from the plan of Nature is readily seen from the actual attachment of the wrist to the forearm as just described.

As Galen thought only casually about this joint, so too it is no surprise that he did not explicate the two rather loose joints in man by which the radius is articulated to the ulna. The ulna admits the radius at the upper joint while the ulna is taken in by the radius at the lower.[148] In the same way, ✷ he neglected the cartilage that originates from the radius and separates the ulna from the wrist and almost hands over the whole wrist joint to the radius.

But as he passed over this cartilage, he also seems to have considered none of the others that in addition to cartilages attached to bones like a crust in smooth joints and by their softness and durability making the bones' joints resistant to injuries, are arrayed in certain joints just as I find the cartilage now mentioned between the wrist and the ulna. Such cartilage is smooth on each side and is attached to the articulated bones only by nearby ligaments. You will find such cartilages in the articulations of the lower jaw to the upper, and in the attachments of the clavicles to the breast bone; so too in the connection of the clavicle to the top of the shoulder or upper process of the scapula.

The wrist bones are not so hard and dense that they are not also visibly porous and filled with marrow, no less than the epiphyses of bones or other parts of bones located outside the large cavities. Though Galen wrote at length about the digital bones, he did not need to add that the digital bones have the same shape as the digits. For although the metacarpal and metatarsal bones are more or less round lengthwise, and can be compared to the digits in that way anywhere, nevertheless as the digital bones appear convex and rounded externally, so too they are compressed internally and smooth in such a

---

[148] As printed in both Basel and Venice editions, *superiori quidem articulo radius ulnam excipit, inferiori autem radius ulnam admittit,* the text is garbled and is corrected in our translation based upon what Vesalius wrote in *Fabrica* I ch. 24 (p. 113): *supra … ulna radium excipit, infra vero a radio ulna excipitur.*

way that with the tendons attached to their inner side (which are like half of a smoothly rounded shape) are only rounded.

I also do not know whether it was Galen's opinion as he wrote it, that all the digital joints are alike and the internode of the previous bone simply enters the socket of the bone that comes next, just as if he wished it said that the attachment of digits is accomplished by enarthrosis. There is in fact a large difference in the joints of the digits. As their motion is also variant and distinct, Nature's great foresight ✳ is provided to them. We flex and extend the first internodes of the four digits, and we move each one to the sides. The second and third internodes experience only extension and flexion and have no common share with lateral motion. It was sufficient for the separation and adduction of the digits that the first joints could be moved to the sides and the other joints would follow their movement by secondary motion. Meanwhile provision was made for their strength by a more powerful type of articulation. For we know that ginglymus is less prone to dislocation than enarthrosis, and the first digital joints are joined in this way, while the second and third joints, as I have written at length,[149] are constructed by ginglymus.

I am therefore surprised at Galen because in *De usu partium* he attributed insertion of muscles moving the digits laterally only to those bones that are moved to the side by their own motion, just as the Ancients had believed, whose opinion he probably used in writing that work. But in *De anatomicis administrationibus* he gives evidence of poorly observed muscle insertions when he says these insertions are made in all the joints of the digits, no doubt confused by the tendons that are attached to the outer side of the digits from the muscles which in various combinations are responsible for their extension and abduction to the side away from the thumb.

---

[149] See *Fabrica* I ch. 27, "The form of the first, second, and third joints of the four fingers" (p. 124).

But someone could justly argue that this has more to do with the traction of muscles than of bones. Of the great number of ossicles that have gotten their name from the sesame seed which I have mentioned in my work, I cannot be sufficiently amazed that Galen clearly explained the location of scarcely a one, and passed them over no differently than if he had found none or very few while dissecting, as they tend to be removed along with ligaments or tendons.

When Galen considered the sockets of the tibia in which the lower heads of the femur are received and stated that they are hollowed out * in proportion to the heads, it is surprising he did not notice how casually and shallowly Nature carved out those sockets if they are to be compared with the large, protuberant heads of the femur; in the process she prepared certain cartilages as a rare and singular benefit and not without good reason, greatly increasing the capacity of the sockets. Galen appears in his examination of the knee joint not to have observed those cartilages, which are smooth above and below and are placed like a crescent moon around the sockets: thick on the outside and gradually ending in a kind of edge, by this means handsomely making the sockets deeper, as I explained carefully in the first book of *De humani corporis fabrica*.[150] I also stated there that the patella seemed harder and of a different constitution than I thought fitting for others to have assigned the name of a cartilaginous bone.[151]

I was also unable there to pass over the long-winded comparison, in the third book of *De usu partium*, of the human foot with the feet of other animals, where I could in no way agree with Galen: he attributed the longest foot to man, but I said it was the shortest, if by the word "foot" should be understood that which is mostly comparable in the human foot to similar bones in number, shape, joints, muscles,

---

[150] Vesalius mentions this omission of Galen first in *Fabrica* I ch. 2 (p. 4), and adds that these cartilages (the lateral and medial meniscus) "were better known to the Arabs." He details their structure in ch. 31 (pp. 138–19).

[151] See *Fabrica* I ch. 32 in the section titled "The substance of the patella" (p. 142).

tendons, and other things of the sort that we consider in describing the parts. Cats, dogs, my weasels, squirrels, hares, rabbits, wolves, lions, and especially simians have the heel bone, talus, navicular bone, and four bones of the tarsus distinguishable from human bones with very few differences. After these comes the metatarsus and the toes, and the proportion of muscles is great. We therefore need to call the body formed of all these the foot and compare it to the human foot. The foot of quadrupeds begins where Aristotle and where Galen in the third book of *De usu partium* established the joint comparable to the human knee, ✳ where they placed the talus in animals with cloven feet.[152] But the talus of the horse, the ass, and any animal with a single hoof also occurs where it does in cloven-footed animals. In quadrupeds having feet divided into toes, the talus is also in the same place, like the talus in man and comparable in use to the talus that cloven-footed animals present to view though they do not correspond very much to each other in appearance.

So it is that the foot of four-footed animals begins from that joint;[153] how much it exceeds the human foot in length when compared with it, is most evident to everyone who is unwilling because of an unduly stubborn devotion that we have for authors to deceive their own reason. I know how many opinions of Aristotle fail as a result of this observation as soon as one learns from me that the femur in quadrupeds is proportional to ours; it is articulated by its round head with the hip bone and attached by two heads to the tibia. Quadrupeds have their tibia articulated to the femur just as in man, and the patella placed in front of that joint as in the human knee. But the femur with its hip joint as well as the knee are seen in certain animals to lie hidden in the trunk of the body. That is probably how Aristotle went wrong here (as I try to show elsewhere concerning the legs of birds).

108

---

[152] Specific references are given in *Fabrica* I ch. 4 (p. 15). See Galen *De usu partium* 3.179.11, Aristotle *De incessu animalium* 711a.

[153] *Viz.* the ankle joint or the joint containing the talus.

In all quadrupeds the tibia is articulated with the talus, beneath which is the heel bone in all of them. The joint here has no other motion than we know it performs in man, flexion and extension. In animals whose foot separates into toes, everything having to do with the movement and succession of bones (but not the number from the sides) comes after this joint, more or less as it does in humans. In animals with a single or cloven hoof, though the series of bones is not the same as in humans, nevertheless whatever joints come after the talus all the way to the toenails have the same ⁕ motion (so to speak) as the joints of human feet in flexion and extension.

In the front feet of quadrupeds, where Aristotle and Galen (in the third book of *De usu partium*) say there is a joint which they would have differ from our elbow, and declare there is a difference between man and quadrupeds, this joint is like the one we have between the carpus and forearm, corresponding in motion and more or less in formation. This is the joint by which the radius of quadrupeds is attached to their carpus. The elbow joint is situated higher; in flexion, extension, and the form of the articulation it corresponds to the joint of the ulna to the humerus in humans. The humerus of quadrupeds also makes an articulation with the scapula that is the same as ours.

In the fingers and other bones of the hands, all are the same so far as we can compare motion. How much these observations conflict with Galen, Aristotle, and everyone who has written often about the locomotion of animals and the erection, sitting, the motion of ankles, and many such things in humans, and how much my statements overthrow their opinions may easily be pondered by anyone who has now been alerted and is held by a desire to know the truth. I do not need to set them forth in detail and at length; I have undertaken only to bring to light a few inaccurate descriptions and observations of Galen. It is surprising that he was so busy making fun of Euripides[154] that he agreed with the opinion of Aristotle. It is certain

---

[154] The point is that humans do in fact sometimes have bone structures like those of horses. In his chapter on the bones of the foot in *Fabrica* I ch. 33, Vesalius wrote "So

that Galen dissected the feet of simians at least and therefore, to mention nothing else, he should have observed that the ape's foot is much longer than man's.

But now an end must be put to my remarks about bones, if I previously stated that Galen was forgetful about the construction of the foot and did not notice that it has one bone fewer than the hand: in the second book of his commentaries on Hippocrates' book *De fracturis* he records the same number of bones in the hand and the foot, not considering that he counted the cubelike bone [*os cuboideum*] ✳ resembling a die twice. First he counted it as a peculiar bone like the heelbone, the talus, and the navicular, and he has this cuboid bone as one of four that make up the upper set of carpal bones. After that he compares the four tarsal bones to the four lower bones of the carpus, not bearing in mind that the fourth and outermost bone of the tarsus is the one that we say resembles the cube or die.

110

Several inaccurate descriptions taken from the account of muscles and ligaments

The definition of a muscle as it appears expertly and handsomely was established by Galen for a person not performing a dissection or who is casually thinking about the nature of muscles. I was not able

---

profusely does [Galen] distinguish man from the other animals in the composition of the bones, and so much trouble does he take to reason why man stands erect and sits, being more occupied in making fun of Euripides than in looking at bones." The reference to Euripides (instead of Pindar) is an error by Vesalius. The reference is probably to a passage near the beginning of the third Book of Galen's *De usu partium* (3.169.15 ff.), where the author comments on Pindar's belief in the myth of the centaurs: "But we who are concerned with truth rather than legends know well that the substance of a man is utterly unable to mingle with that of a horse." See Pindar's Second Pythian Ode, 44–8. By the time he published the 1555 edition of the *Fabrica* Vesalius corrected his error.

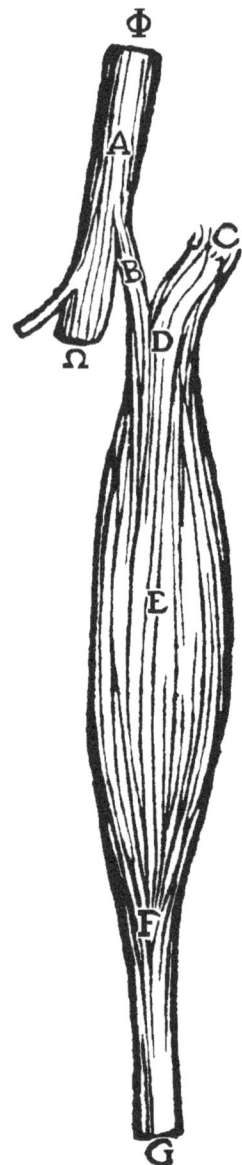

to support that definition, and have had to establish another, which is in the second book of my work *De humani corporis fabrica.*

Sample muscle from the beginning of Fabrica II ch. 2. Between Φ and Ω is a nerve segment (A) from which a twig (B) innervates the muscle. C is the muscle's origin, G its insertion as a tendon. The belly is labeled E.

I have so far observed no muscle in the entire body for which Galen's description is perfectly adequate to fit the nature equally well of both ligament and nerve, or to set some exact distinction between a nerve and a ligament.[155] This is not to say that an equal portion of ligament and nerve can not enter the makeup of muscles. Though Galen is seen to indicate that more often, he elsewhere judges that a portion of nerve is less than a ligament. The fact that a muscle is not intermediate by nature, or the tendon inserted in the back of the heel, will prove what is by itself much thicker than the roots of four sinews distributed into the femur if one imagines them to have combined into one body.

How many other tendons are there, I ask, which are inserted into the tibia, the fibula, and the bones of

---

[155] The context requires that we understand "tendon" for *nervus* in these paragraphs. In the Latin, *medium quid inter nervum & ligamentum* must mean a medium of distinction between tendon and ligament. Lat. *nervus* can mean sinew, muscle, tendon, or nerve. Similarly Gk. νεῦρον can mean sinew or tendon and the pl. νεῦρα can mean nerves as organs of sensation. The English expression "strain every nerve" preserves the ancient ambiguity. The resulting confusion adds no small obscurity to these paragraphs, where *nervus* can also be a ligament but *nervosus* regularly means "sinewy." In modern anatomy a ligament connects bones or supports viscera, while a tendon connects a muscle to its origin or insertion.

the foot, having been brought from the muscles to which it is known the animal force[156] is borne from those four roots? And when Galen applies a nerve to the ligament in the origin of a muscle so that they may be blended together, he believes both are divided into branches; separated in turn into several, they eventually join gradually into one and make up a tendon. Who has not noticed that such a nerve combination cannot be made from muscles that at their beginning are wider than a tendon, ✳ and do not show their fibers more scattered anywhere than at their origin? The muscles responsible for motions of the femur that originate from the iliac bone are of the same type, in addition to many movers of the humerus. The intercostal muscles are no less wide at their insertion than in their origin, and throughout their course. The muscles that we call the transverse and oblique abdominals are wider where they become tendinous than at their origins and beginnings.

111

I need not mention that nerves sometimes form a series that is nearly identical with the veins and arteries in the muscles, and sometimes above the muscles when these are small and thin. And if you study this series you will sometimes observe that a nerve runs to the muscle far from the muscle's head, without doubt for the beginnings of the muscle. For it is clear enough that nerves are presented to the heads of muscles that I have said are in charge of movements of the femur and humerus, but not in the way that Galen indicated, since Nature did not fail to prevent entries of nerves in a direction contrary to the course of a muscle's contraction and movement.

Again, if you compare a boiled tendon with a nerve [ligament] cooked at the same time and cut both transversely, you will discover that the tendon differs from the ligament in density, continuity, and substance; but the nerve [ligament] will look like a cord formed from

---

[156] Vesalius' *animalis vis* or *animalis spiritus*, Galen's psychic pneuma, was thought to be produced in the brain and sent through the invisible lumen of the nerves, responsible for motion and sensation.

several strands and resembles the tendon in neither form nor substance. The tendons of muscles are not thicker because of the blending of nerves with their heads or origins as Galen thought where he tried to show the blending in of nerves.

I will not discuss muscles that originate with a fleshy beginning; they have an origin nearly three times larger than their insertion, as are nearly all that are responsible for motions of the femur. Let us consider muscles that are brought forth more or less sinewy and do not immediately become fleshy. Let us then see whether the first, second, or fourth muscles moving the foot make up the greater part of a sinewy origin $*$ than the tendon of all of them, inserted in the heel, has a thickness. And who does not see more fibers and sinewy origins in the muscle raising the arm than are seen at its insertion into the humerus?

It should also be surprising that Galen more than once says that no muscles moving bones are without tendons, though in many bones it is otherwise. For example, there are no tendons on the intercostal muscles, nor in the quadrangular muscle pulling the scapula toward the back, the muscle adducting the thumb to the index finger, the lesser of the muscles flexing the first bone of the thumb, and many of this kind. Though one may on rare occasions have observed the nature of a tendon as I have described it, and reflected that flesh not only was created for separation of the fibers in the muscles as Galen believes, but must be considered the principal substance of the muscle, he will certainly have no difficulty understanding that the Maker of things first gave tendons to muscles when the muscles occupied a place along whose length they could not anywhere be made of flesh. When they are contained in a narrow place and are not extended in length, they have no tendons, or they have long or short tendons according to their location.

There is a similar place where Galen states a number of times that no muscle originates except from a bone. He should have noticed at that time that the eye muscles do not originate from bones, especially the one[157] surrounded by the first six muscles. Moreover, the muscles adducting the four digits of the hand to the thumb take their beginning from the membranes that wrap the tendons flexing the third bones of those fingers. The round muscles of the anus and the neck of the bladder also do not start from a bone.

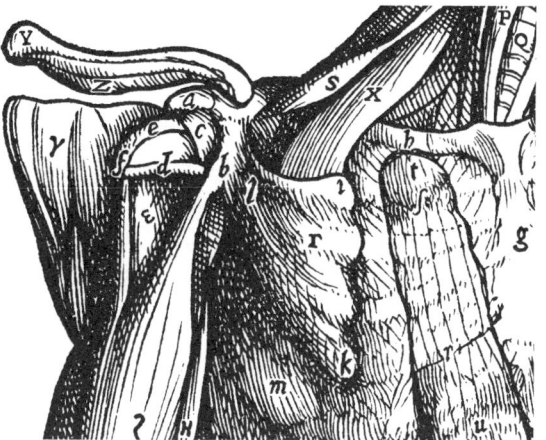

In this detail from the fifth figure in Fabrica II, the triangular pectoralis minor muscle is labeled Γ. Two of its origins on the rib cage are labeled i and k and its insertion into the scapula is marked l. The deltoid muscle, partially removed, is marked γ.

I am equally surprised that Galen counts eight common muscles of the larynx, since he describes only six. I believe he wanted to add to the count those that are inserted into the covering of the larynx from the hyoid bone and which I found move it upward, if even a slight mention of them ⁎ had occurred in Galen and he had not said everywhere that the common laryngeal muscles are inserted into its first [thyroid] cartilage, the one resembling a shield, and that they all go there.

I shall state elsewhere that Galen did not rightly understand the function of the muscle [*m. pectoralis minor*] that adducts the scapula to the chest[158] and is inserted into the inner process of the scapula.

---

157   The *retractor bulbi*. But this is not a human muscle. Vesalius' account of that muscle in *Fabrica* II ch. 11 (pp. 240–21) is based in part on Galen's description in *De usu partium* 3.792.12ff. (May 1968, 483f.). To judge from his descriptions and illustrations of the eye in *Fabrica* II ch. 11 and *Fabrica* VII ch. 14, there is no evidence that Vesalius ever dissected a human eye, and violated his own cardinal rules against trusting descriptions found in books and basing human anatomy upon the anatomy of animals.

158   See *Fabrica* II ch. 23 (p. 263): "it does not, as Galen maintains, move the arm to the upper chest; the fact is that it was made for the scapula to draw it to the chest, as

He also describes it badly, not only in other areas but especially at its origin. Though Galen more than once says otherwise, it never takes its beginning from the second rib of the thorax and the three that come after it at the point where their cartilages are attached to the chest bone, or for that matter from the chest bone itself. Rather, those ribs provide a beginning for the present muscle before they end in cartilages. Since Galen did not understand the origin of this muscle in apes or in humans, it is no surprise that he put its insertion in the shoulder joint, though in fact it makes its entire insertion in man into the inner process of the scapula. In caudate apes, where that process is extremely small and scarcely projects, a portion of the insertion is seen extending to the ligament surrounding the shoulder joint, though because of this insertion the muscle cannot perform the function that Galen wrongly ascribes to it.

In the description in Galen of the second muscle [*m. deltoideus*] moving the arm, resembling the appearance of a Δ, it is seen that several muscles are mentioned beneath it. As the muscle which I consider the only one resembling a Δ is formed from many that lie upon it, so too we know that the first three that move the femur and take their beginning in a continuous succession from the iliac bone are layered on each other and mutually attached. In these descriptions Galen is no more truthful than he is consistent with himself in the thirteenth book of *De usu partium* and the fifth of *De anatomicis administrationibus*.[159] However that may be, you will find no muscle in this place besides the one I have just mentioned. Nor do I believe that it receives aid ✳ from any other muscle in elevation of the arm.

its origin, course, and the insertion of its tendon show clearer than light, particularly in man." Vesalius cites Galen *De anat. adm.* 2.480.16: "A third muscle remains which becomes visible when [the *pectoralis major*] is removed. It, too, springs from the sternum, [but] at its junction with ribs 2 to 6. It is the highest that adducts the humerus." (tr. Singer 1956, 122).

159 As argued in *Fabrica* II ch. 23 (p. 265). See Galen *De usu partium* 4.135.1–5 (tr. May 1968, 616) and *De anat. adm.* 2.489.14–16 (tr. Singer 1956, 126).

For if I should assign to this place a muscle that like a part of the muscle occupying the hollow part of the scapula fills the hollow in the convexity of the scapula and the area that stands between the upper rib of the scapula and its spine, I would not be able to give it the smallest space in which to make contact with the deltoid muscle. So far would the scapula be from attaching to the deltoid by a significant connection: that is what all Galen's descriptions are seen especially to lack.

As Galen occupied himself in vain establishing other muscles here, so too he neglected the insertion of the deltoid muscle, which he said takes place directly downward into the humerus, writing that its insertion is like that of the muscle adducting the arm to the chest. The fact is that the great craft of Nature fashions a more or less transverse insertion that angles quite obliquely downward to the rear. From the motion that we make when erect, moving the arm to the forehead, nose, ear, and then far behind the occiput in a kind of semicircle by means of the deltoid muscle, it is perfectly clear how necessary it was that it make the insertion that I describe and is seen in dissection, and not the one that Galen wanted, which would be taken straight downward along the length of the humerus.

These things that I have said I would like you to learn about caudate apes as well as humans, for they are like us in the present muscle while dogs and other animals without a clavicle are very different. The fourth muscle [*m. latissimus dorsi*] moving the arm, which originates from tips of the vertebral spines starting with the spine of the sixth thoracic vertebra as far as the middle of the sacrum, is triangular in shape and moves the arm downward; since the base of the scapula is covered at its higher rib by this muscle's lower angle, it makes no insertion into the scapula, nor is it attached to the muscles lying beneath it there beyond its entire width, though it appeared to Galen that this muscle was attached to the scapula with a large and significant grip ✳ where he invented muscles pulling the scapula downward.

115

137

The insertion is not the only thing Galen attributed wrongly to this muscle: he also believed it took its beginning lower down when he wrote that it begins where the lower of the posterior muscles moving the scapula takes its end. I suppose that is what he called the lower part of the muscle that we consider the second [*m. trapezius*] of the muscles moving the scapula and compare to a monk's hood. This part faces downward more than the beginning of the fourth muscle moving the arm [*m. latissimus dorsi*]

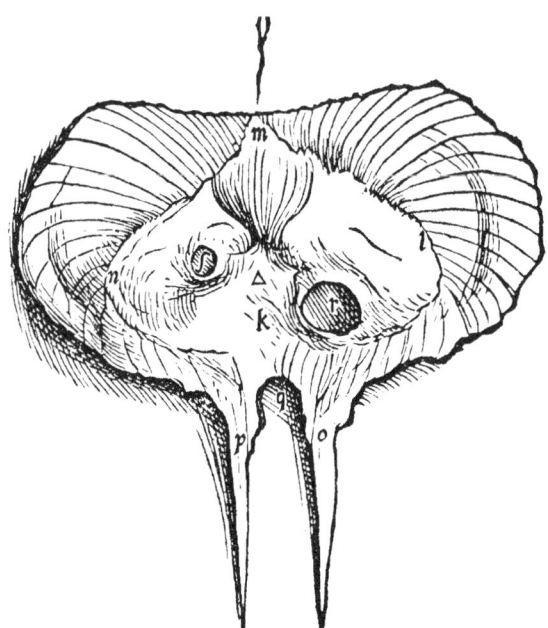

Detail of the 7th Table of Muscles in Fabrica II showing the diaphragm and its openings. The elongated s marks the foramen of the vena cava; r identifies the hiatus oesophageus that admits the gullet into the abdomen, and q is the hiatus aorticus through which the aorta descends along the vertebrae.

because it has that part spread upon itself for a noteworthy interval.

I consider the muscle [*m. iliocostalis thoracis*] running along the muscles of the back lengthwise on the thorax and taking its origin from the fleshy parts of the loins to be the fourth of the muscles moving the human thorax. It does not consist only of fleshy parts (though it seemed otherwise to Galen) since it makes such handsome tendinous insertions into tubercles of the ribs provided for it.

In his description of the foramina of the transverse septum, I am in disagreement with Galen when I attribute to the esophagus, the path by which food and drink are delivered into the stomach, a peculiar foramen separate from the area [*hiatus aorticus*] of the trans-verse septum where it yields to the vertebral bodies and the great artery on its way to the loins; Hippocrates and Galen count them as a single foramen and place it above the eleventh vertebra of the thorax, arguing that it also provides a path for the gullet. The upper orifice of the stomach does not stand directly opposite the twelfth thoracic vertebra, or even if one wishes, the eleventh. It is therefore obvious that the esophagus could not descend that far. For that reason it has a peculiar foramen [*hiatus oesophageus*] by which it accesses the stomach; this foramen is not only located above the path of the great artery, but it also inclines slightly to the left side of the septum as does the upper orifice of the stomach.[160]

Concerning the orifices of the stomach, ✳ there will be sev-eral matters to be mentioned in what follows. For the present it will suffice to have given notice that the esophagus and offshoots of the sixth pair of cerebral nerves that run alongside it have a separate fora-men, as does the vena cava. My careful description of muscles moving the back, and the very perfunctory enumeration from Galen that is scarcely mentioned, easily prove Galen's negligence regarding those muscles. To pass over other arguments, he missed four muscles mov-ing the back which he cut off as if they were parts of other muscles. He omitted two that extend from the end of the posterior part of the sacrum to the vertebra [12th thoracic] that is supported above and below, are attached to the intermediate vertebral spines, and are thin at their beginning and end but thick in the middle. He neglected two others that extend from the thoracic vertebra that we say is received above and below, and run to the neck; they occupy the spines of ver-tebrae along which they are stretched and are very strongly attached to them, being not so different from the recently mentioned muscles except that they are as much smaller as the thoracic vertebrae are

---

[160] For more on Vesalius' critique of Galen on this point, see *Fabrica* II ch. 25 in the section titled "Foramina of the septum" (p. 291).

more obscurely moved than the lumbar vertebrae. As soon as these muscles grasp their spines and pull away from each other we experience motions of the back. Galen did not notice those four muscles; but how perfunctorily he described all the others one may learn from his books.

Among the ways related to vertebral ligaments in which we may miss Galen's customary diligence in dissections, not the least appears to me to be that he disagrees with the old experts in dissection who we believe from Galen's books were thoroughly trained in the anatomy of bodies: he disagrees with the opinion in which they affirmed that the vertebral bodies are connected by the cartilage that lies between them. We observe between the vertebral bodies a certain mucous, soft, and fibrous cartilage than which I should think no other body can better deserve the name of cartilaginous ligament. I can think of nothing more certain ☀ in the whole body than that these cartilages, or rather cartilaginous ligaments, join together the vertebral bodies.

117   I therefore wonder at the invention of Galen where besides those cartilaginous ligaments he says that a process of some third wrapping of the dorsal medulla (which is probably nothing other than the vertebral membrane corresponding to the one that we know covers the other bones as well) is brought between the vertebral bodies, and that this, rather than the cartilaginous ligaments, is how the vertebrae are joined. The third wrapping of the dorsal medulla is not unknown to me: but I know perfectly well that it sends no processes between the vertebral bodies.

But no one who has taken the trouble to study the boiled vertebrae of a lamb, a kid, or a calf could miss Galen's calumnies against the Ancients. By this means he should examine three (as I would say) sets of cartilage between two vertebral bodies, separated by two ossicles. There will then appear an epiphysis of each vertebra and a cartilage by means of which the epiphysis blends with the remaining bone, and finally the cartilage or cartilaginous ligament placed between the two epiphyses. Thus when you decide to separate the vertebrae from each other or break them apart, you will notice that they do not separate between the epiphyses but between the epiphysis and the remaining bone, and the distinction will be clearly apparent between cartilaginous ligament and the cartilage otherwise provided for the attachment of bones.

When Galen tries so hard to describe the tendon concealed under the skin of the hand and then to explain how difficult it is to separate from the skin, and how it is coterminous with the skin, he should have considered how great a quantity of globular, rather hard fat comes between the skin and the tendon. As it was necessary for the tendons flexing the second bone of the four fingers to be inserted into that bone, and before their insertion they would have to be divided by a long slit in order to transmit tendons running to the third joint, ✳ Galen should also have observed that this slit or opening is not made above the second digital bone but before the tendons have passed 118 the first bone. When the tendons flexing the third digital bones rest upon the first bone, they are not hidden; indeed, they remain perfectly round just as they do throughout their course, though Galen said otherwise, having written elsewhere (still more wrongly) that they are inserted in the first bone. Moreover, the tendons flexing the second bone [*m. flexor digitorum superficialis, tendo*] are not inserted into the sides of those bones as Galen's opinion holds, but lie hidden and place themselves on each side beneath the tendon going to the third bone, and the parts coterminous with it on each side are implanted on the inner [palmar] surface of the bone.[161]

---

[161] This disagreement with Galen is aired on p. 306 of the 1543 *Fabrica* in the section titled "Transit of the tendons of the second muscle through the tendons of the first," where a marginal note cites Bk. 1 of *De anat. adm.*; see 2.250.16–18: "as each passes over the former larger tendon, each splits in two, encircles the tendon lying under it, and is attached to the sides of the second phalanx." (tr. Singer 1956, 16). Cf. *De usu partium*, where the divided slips of the superficial tendon are inserted into "the inner parts" (τοῖς ἐντὸς μέρεσι, 3.59.10, May 1968, 97) of the head of the second phalanx. Modern anatomy sides with Galen against Vesalius: these slips are regularly described as inserted into the sides of the second phalanx.

Although after writing *Du usu partium* Galen prides himself while writing the second book of *De anatomicis administrationibus* that he has discovered ten muscles by which the first joints of the fingers are flexed, he always overlooks the three that serve flexion of the thumb. For in addition to the muscle extended from the forearm that performs flexion of the third joint of the thumb, we find five that also aid flexion of the thumb, two of which preside over flexion of the first bone while three flex the second. Because Galen counts ten muscles flexing the first bones, we will acknowledge that two of the total were known to him that serve flexion of the first bone of the thumb; we make this guess because he did not specify the muscles' shape, location, or nature. We shall therefore say that the muscles that are inserted in order as if on the course of the life line from the palm into the second joint of the thumb and are authors of its flexion according to the excellent design of nature, were not observed by Galen. In the same way, Galen passed over the peculiar tendon [*m. extensor pollicis longus, tendo*] inserted in the root of the first bone of the thumb on its posterior side and not climbing higher like the tendons that are provided to the outer side of the thumb. That tendon comes from a portion of the third origin of * those that originate one after another along the length of the ulna, as is written in my book. This third origin is the one that we observe during dissection is more coterminous with the wrist, and which Galen wrongly said originates along the entire length of the ulna. Though there are about four roots of tendons to the fingers on the outside of the arm, we sometimes see a different series.

I will not cite here the place where Galen adds that the four fingers are extended by one muscle split into four tendons, while the index and middle fingers are abducted sideways from the thumb by one muscle divided into two tendons, and the ring finger and little finger are similarly angled to the outside by the other tendon. For although these occur differently in man, they are not all arranged in the same series always.

However, it is altogether necessary to go against the position of Galen when he teaches that the tendons extending the fingers are

119

attached only to bones, where there are structures of joints when the tendons are inserted equally the whole length of the bone. Also, we always see that the muscle placed in the hollow of the palm that abducts the thumb farthest from the other fingers is not inserted in the first bone, however much Galen taught otherwise. It is implanted in the inner side of the second bone by a rather short tendon, and it is no wonder that Galen described its insertion that way in *De anatomicis administra-*

Detail from the second Table of Muscles in Fabrica II showing the extensor retinaculum, the band-like ligament forming the roof of the carpal tunnel, distinguished into six parts described as follows: "1 marks the ligament common to the radius and the ulna; 2 is the ligament belonging to the ulna; 3, 4, 5, and 6 introduce the four ligaments peculiar to the radius."

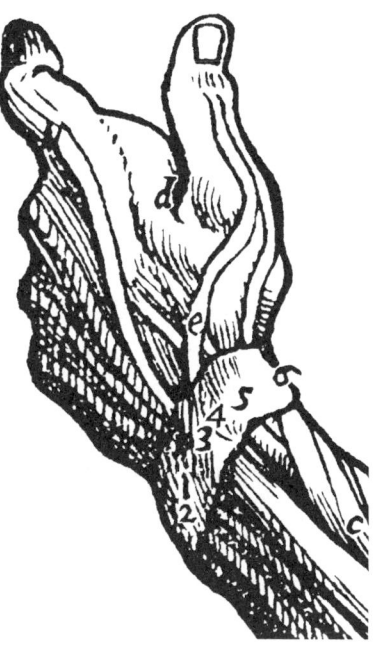

tionibus though when writing *De usu partium* he thought this muscle was the whole fleshy mass placed at the first bone of the thumb and called Venus' mount by chiromancers; but we say it is made up of six muscles.

Similar to this was the opinion of Galen where he thought the fleshy mass placed below the little finger and named after a mountain of the Moon[162] was only the muscle that abducts the little finger farthest to the outside, when in fact the muscle forming the greater portion of this mass flexes the first bone of the little finger to the outside.

As for the insertions of tendons moving the fingers, * these 120 obviously differ from Galen's account, not only in some that have

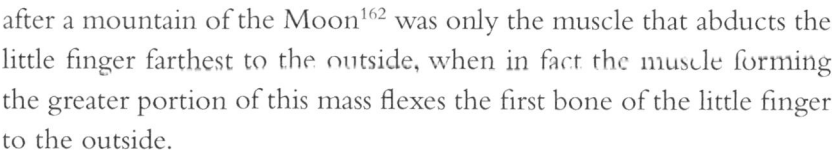

---

[162] The hypothenar eminence, here called the *lunae mons* and *lunae monticulus* in *Fabrica* II ch. 43.

already been mentioned but also in the implantations of those that move the fingers to the sides (for neither I nor the doctors who preceded Galen assign such insertions to joints which do not have their own movement to the sides), and I cannot agree with his opinion in the number of insertions.

From a painstaking study of the grooves in the epiphysis of the radius and inspection of the transverse ligaments at the end of the forearm, it is also generally admitted how much we still find lacking in Galen's descriptions. Besides the fact that he recognized only four of the six and sometimes seven rings that occur there, he also traced a false course for the tendons together with inappropriate grooves and ligaments. For example, when describing tendons of the muscle by which he thought the four fingers are extended, he says they are contained within a common ligament and the groove of the ulna and radius. In fact, a muscle occupies that groove and ligament; it is inserted principally in the little finger, and is held by Galen to be the author of abduction of that finger and the ring finger from the thumb. An elegant groove is prepared in the radius for that muscle, which is split into as many tendons.

I am still more amazed that Galen did not mention the membrane which is tenaciously attached like muscles to the surfaces of the transverse ligament that face the skin, just as if muscles needed to be restrained by that membrane to keep them from being moved out of position. That is especially observed in the forearm and lower leg, while in the thigh (as I have written)[163] that function is performed only by a peculiar muscle. At this point the square muscle[164] pronating the radius covers the inner side of the radius and ulna: but it does not, as Galen teaches, come between those bones.

The muscle [*m. brachioradialis*] originating from the humerus and inserted in the epiphysis of the radius, causing it to supinate, is not ✳

---

[163]   In ch. 53, *Fabrica* II (p. 334, misnumbered 234): this is the *m. tensor fasciae latae*, Vesalius' sixth muscle moving the tibia.

[164]   *M. pronator quadratus*, Vesalius' first of four muscles specifically moving the radius: *Fabrica* II ch. 45 (p. 315, misnumbered 215).

the longest of all the muscles moving the forearm[165] if we compare to it the muscle flexing the third joints of the fingers or the one held to be the author of extension of the fingers, or the one that is implanted by the two-horned tendon into the metacarpal bones and takes its origin from the humerus immediately under Galen's longest muscle.

Galen little noticed that the inner beginning of the anterior muscle flexing the forearm [*m. biceps brachii, caput breve*] is much wider than the outer, taking the shape of a rounded tendon. Not only is it wider, but it originates with a substance that is part fleshy and part tendinous, with the result that the fleshy part seemed like a separate muscle when I first read in Galen that the inner beginning was more slender than the outer.

We do not know to what extent Galen understood the muscles of the penis and those that occur in the wrapping of the testicles and seminal vessels that comes from the peritoneum. There is never an actual mention of them, or of the muscles of the tongue. But if only because Oribasius[166] and the Arabs mentioned them I can easily understand that he wrote about them in the books that have been lost to us in the damage of the ages.

From the muscles moving the femur as mentioned in *De anatomicis administrationibus*, I shall not attempt to augment the list of inaccurate descriptions because I can find no end of them in that work. I also can scarcely understand some of them therein, and I think his examples egregiously mendacious, except in his account of the muscles moving the lower leg, which I considered in both *De usu partium* and *De anatomicis administrationibus* when writing the 53rd chapter of

---

[165] As pointed out in *Fabrica* II ch. 45; the brachioradialis is Vesalius' second muscle moving the radius. Galen had said "it is the longest not only of the muscles that move the radius but also of all the other muscles of the forearm." (*De usu partium* 3.113.17–19, tr. May 1968, 127).

[166] Physician to the emperor Julian, Oribasius was a 4th-century compiler of Galenic teaching. His *Collectiones medicae* preserve a large number of excerpts from ancient writers. Through Syriac and Arabic translations he was a principal conduit of Greek medical learning to the Islamic world.

Book II of *De humani corporis fabrica*. For anyone reading that chapter, the aforementioned list of errors will be summarized and increased not only in muscles omitted and overlooked, but also in the origin and insertion of the ones he knew and in others that were improperly described, in the same way as in my account of the muscles moving the foot and its digits.[167]

122

Vesalius' picture of the portal vein system from Fabrica III chapter 5. The liver is at the top, with offshoots of the portal vein marked by five A's. The main trunk of the portal vein is labeled B. This figure of the portal system as a whole illustrates how well the plant metaphor worked for Vesalius.

I could therefore justly seem to you ridiculous and too much of a drudge if I thought I should give you an enumeration of all the matters of dispute in my book, especially since out of what has been so far mentioned it should be apparent to anyone of sound mind that something has been misreported by Galen in his description of muscles. Sylvius, on the other hand, has been objecting that Galen stated nothing whatever less than perfectly.

---

[167] Chapter 59 of *Fabrica* II.

Some false descriptions gathered from the account of veins and arteries

But let us see whether nothing false or erroneous was stated by Galen in the other organs as well, starting with the veins.

The beginning of the vena cava does not take place in the middle of the convexity of the liver in the same form as the beginning of the portal vein, which we know arises even more in the back of the liver than the front, and faces right rather than left. We see one trunk in the hollow of the liver which we call the beginning of the portal vein because it is combined out of many countless branches

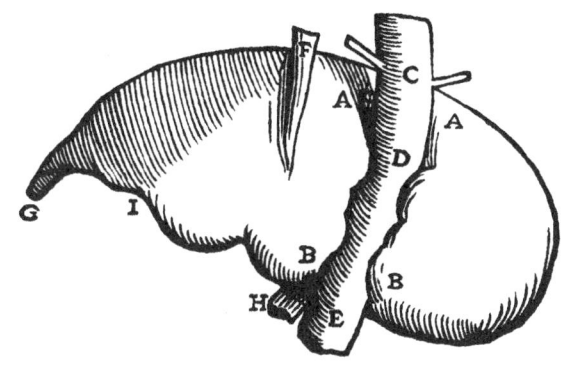

The vena cava (C, D, E) embedded in the back of the liver, as illustrated in Fabrica III ch. 6 and Book V fig. 18.

distributed through the liver; or more accurately, this trunk distributes that series of branches variously into the liver. In animals, the first of all is split into as many branches as there are fibers: five or six lobes are found, and one branch is sent equally to each of them. In man, whose liver is formed of a continuous and simple body, the stem of the portal vein is generally divided first into two branches, and these are soon distributed into other offshoots; those offshoots are scattered into countless twigs and subdivisions so they can travel everywhere through the lower part of the liver.

The stem of the vena cava does not rise with a similar collection of branches, but is located in the back of the liver and impresses a groove there where it is seen surrounded by the liver on its front and sides but on its posterior free from the substance of the liver, not surrounded by the flesh of the liver. Thus even when placed in this spot the stem of the vena cava easily shows how well it agrees with Galen's description regarding its position. ✳

123   However, the vena cava differs in its series of branches more than in its location. From the front of its stem, where it stands in the liver, two branches are brought forth not far from each other but separated transversely; these are then split into several offshoots, and when those offshoots have been additionally distributed into a series of countless twigs they run off through the upper part of the liver, extending alongside and lying upon branches of the portal vein.

Besides the two branches originating from the front of the stem of the vena cava, two or three twigs are also presented to the liver in its descent along the back of the liver, perhaps ten times smaller in their orifice than the large branches and distributed by a small, short entry through the substance of the liver.

However that may be, if you gather together all the mouths of branches and twigs distributed from the stem of the vena cava into the liver, your total will certainly not come to half the volume of that stem: so far is it from being possible to say truthfully that the vena cava is made up out of those branches.[168]

These matters should not much bother the viewer in dissections if he is aware that the fabric of the body is so to speak diametrically opposed to what Galen described when he showed the vena cava originating from the liver. No one can ever set aside his loyalty and examine these matters without wondering how Galen could have placed such trust in his own imagination and dared to contradict Aristotle and other weighty authors in this part of the body, teaching later generations what is utterly at odds with what we observe in bodies.

---

[168]   This recapitulates Vesalius' refutation of Galen's argument in *De placitis Hippocratis et Platonis* that the liver is the source of the veins as the heart is of the arteries and the brain of the nerves. See *Fabrica* III ch. 6.

Just as Galen's description of the origin of the vena cava is quite discrepant from what we observe when dissecting, so too it is extremely false and unworthy of an anatomist that the beginning of the vena cava (as Galen wrote) forms a single stalk like the great artery originating from the base of the heart, and is then like the great artery split into two trunks, one traveling upward and the other down. Instead, the stalk of the vena cava in this part has nothing in common with the great artery. ⁕

It travels vertically along the back of the liver just as if taken from the heart beneath the liver, and goes to the lower body; it also provides branches to the liver (as had been said) from its anterior side. The great artery, on the other hand, proceeds upward for a while from its beginning as a single stalk and is then divided into two unequal trunks, the larger of which turns downward and is distributed to the arteries that are beneath the heart, while the lesser trunk goes to the upper body. I never thought that Galen indicated the similitude in distribution of the vena cava and the great artery (except in actual function, so to speak), unless to oppose Aristotle[169] he had taken his argument from the pattern of distribution (also against what Hippocrates said where he taught that the vein takes a direct course) where he said the vena cava does not originate from the heart. How effective that argument is, will perhaps be indicated somewhere below.

Here, though, it suffices to have made the point that Galen's opinion is quite wrong, as is the description where he set it down that a single stem of the vena cava originates from the middle of the convexity of the liver and is then like the great artery divided into two trunks one of which passes beneath the liver and is presented to the

---

[169] For Aristotle's axiom that the heart is the source of all blood vessels (ἡ δὲ καρδία τῶν φλεβῶν ἀρχή) and the arguments defending it, see *PA*665b15 ff. A fuller version of Vesalius' argument against Galen on this point is in *Fabrica* III ch. 6 in the section titled "Galen's arguments against Aristotle on the origin of the vena cava are not all divinely inspired" (p. 375, misnumbered 275).

parts lying beneath it while the other permeates the transverse septum and provides nutriment to the organs above it.

Just as he taught us a false partition of the vena cava, so too it is no surprise that he said the portion of the vena cava visible in the loins beneath the liver is much larger than the one to which the transverse septum provides a path, taking

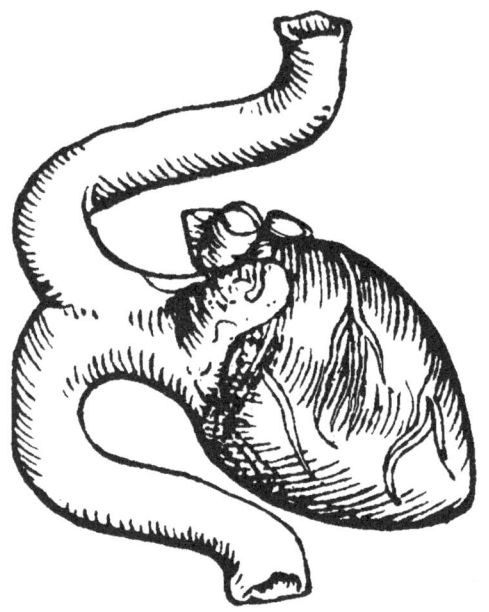

This imaginary illustration from Fabrica III ch. 6 shows how the vena cava would appear next to the heart according to Galen's description: "In this figure I have drawn the arrangement of the vena cava that would inevitably result if it were split into two trunks at the right side of the heart. To observe more precisely Galen's argument, which he repeats so often and uses instead of the best demonstration, compare the present figure with some of those in the sixth Book" (p. 375, misnumbered 275).

his reasoning from the parts which each nourishes: Galen affirms that the trunk going lower down is larger because it must provide nourishment to more parts than the one that goes higher.[170] As this occasion is not ineffective in explaining the rationale of distribution, so too Galen should have considered carefully whether the part of the vena cava below the liver supplies nutriment to more and larger parts than the one that * stands above it, so that he would then draw the right conclusion from his syllogism. It should be gathered from Galen's reasoning

125

---

[170] See *Fabrica* III ch. 7 in the section titled "The part of the vena cava going through the transverse septum [*diaphragma*] is not smaller than the part below the liver."

from the nutrition obtained that the part of the vena cava that permeates the transverse septum is greater than the lower part. For the gall bladder, stomach, omentum, spleen, and all the intestines with the mesentery take nourishment from the vena cava below the liver and not a series of branches; only the kidneys, genitals, the other bladder, the lower part of the abdomen, the loins, and the legs take their nourishment from a series of branches. The septum itself and the entire thorax as far as the first lumbar vertebrae together with the upper region of the abdomen are nourished by the portion of the vena cava crossing the transverse septum, as we know from the offshoots of the azygos vein and then from the veins passing beneath the breastbone and descending almost to the navel. To these are added the arms, neck, and head, where the mass of the brain must be considered, and the large number of vessels going to the brain.

Now the lung also presents itself, the largest member of the whole body, though it is otherwise if it lacks blood. Were it not for the lung, blood would not be brought into the heart, but so all the arteries would eventually be filled by it. Thus the greatest proportion must be made up of the blood which is contained in the arteries. It is therefore evident that the portion of the vena cava visible between the heart and the liver should be larger than that which Galen knew stands at the lumbar vertebrae. Dissection also supports this thinking if a person investigating these matters cuts open the vena cava lengthwise or at least does not in the process of dissection force more blood from movement of the liver or the heart or rather its septum into the part of the vena cava that extends to the loins than he forces into the part that stands in the septum. When the vena cava passes through the transverse septum, in the part of its course that is seen between this point and the heart, it presents no offshoots to the membranes that divide up the thorax, though Galen taught otherwise, since it does not make contact with those membranes; and it would have been ridiculous for such small twigs as they require to travel supported by no membrane. *

For this reason Nature provided the membranes chiefly with   126
offshoots taking origin from those veins which run beneath the

breastbone, and exclusively in humans from veins that run out along the nerves of the transverse septum from the throat to the septum.

Moreover, in what ways shall we apply Galen's description to the body, a description in which he explains the distribution of the vena cava to the heart, sometimes taking a kind of offshoot from it into the heart, sometimes explaining it in some other way that makes the orifice by which the vena cava is joined to the heart twice as large as the vein's diameter? This is the same as if from its circumference where it stands beneath the heart and then from its circumference as you would measure it where it passes the right auricle of the heart you would imagine a single circumference in which we combine in one circle the capacity of two circumferences or circles. In this way the amplitudes of the vena cava above and below the auricle of the heart would be two circles which would make up a single circle corresponding to the circumference of the orifice by which the vena cava reaches the heart.

When Galen appears to tell in his book *De venae sectione* about the arterial vein [*truncus pulmonalis*] as if it came out of the vena cava, I do not believe he thought that it was an offshoot of the vena cava because he did not fail to understand otherwise that the substance of the arterial vein does not resemble that of the vena cava: the latter is composed of a single tunic and the former of a double, the inner of which is common to all arteries, as much as five times thicker than the tunic of the vena cava, though on the outside matching the thickness of the tunic of the vena cava. Also, the arterial vein has its own peculiar orifice, like that of the great artery, besides which the substance of the heart between the orifice of the vena cava and the beginning of the arterial vein is at a substantial interval, easily demonstrating that the arterial vein does not originate from the vena cava.

Though Galen affirms elsewhere that two coronary veins are always found, we generally find a single one, ✳ and it is large, not as Galen also says elsewhere, small. Also, though Galen stated in the seventh book of *De anatomicis administrationibus* that the coronary vein arises from the right ventricle of the heart, I do not think he had so briefly considered the planning of the Maker of things in the course of this vein that he perceived it had left the vena cava before it left the right ventricle or had originated before it entered the ventricle, and that its origin is seen under the base of the three membranes that control the orifice of the vena cava.

Sylvius bore it ill that I have written that Galen did not dissect humans and added separately that Galen had viewed only the outer veins of the arm without seeing the inner veins or those hidden deep in the body. I agreed; I have written a few things about the series of veins in the throat and the arms in the place[171] where I cited several reasons out of many why I am persuaded beyond a doubt that Galen had no experience in human dissections. Here I set down an example or two of his inaccurate descriptions, such as the origin of the humeral vein and the series coming from the trunk of the more important and larger axillary vein: in these examples one might justifiably look for Galen's diligence, no less than in certain veins of the legs, and especially in the cerebral vessels, where he mentions scarcely two of the six veins entering the skull on each side. Similarly, of the three arteries entering the skull he explained the distribution of only the one from which he thought the reticular plexus was formed. Thus in describing the sinuses of the hard membrane it should not seem strange if he believed that the chief ones acted only as veins and paid no attention to the arteries draining into them or leading to them, and did not notice that the sinuses also contain the substance of veins and arteries.

Based therefore upon my description of cerebral vessels in the fourteenth chapter of Book Three of *De humani corporis fabrica*, ✳ if it is ever compared with Galen's account in the ninth book of *De anatomicis*

128

---

[171] Chapters 7 and 8 of *Fabrica* III, where Vesalius describes the beginning of the humeral vein and the distribution of the humeral and axillary veins in the arm. Vesalius' words here anticipate additions he will make to the 1555 edition, where his critique of Galen regarding these veins is much expanded.

administrationibus many things will present themselves in which it will be quite easy to observe that Galen fell short of the true distribution of the veins and arteries. If I were to explain all of them, the heap of Galen's untrue descriptions would grow too much. Instead, it will suffice to have mentioned a few examples in the distribution of vessels.

Accepted descriptions in the account of nerves which are not quite true

Granted that the opinion in Galen about the olfactory organ (to add something about the nerves) is inconsistent, the view that he holds is sounder where he says that the nervelike processes running in the anterior part of the brain are for the organs of smell. But the view that the organs of smell run from the anterior parts of the cerebral ventricles which end in a point or narrow place is altogether untrue. First, the right and left cerebral ventricles are large, wide, and not compressed where they are closest to the anterior part of the cerebrum and face forward. Second, if one measures the course of the ventricles along the length of the brain and then in the base of the brain, and notices that these processes thought to exist for the organs of smell take their origin more or less half way along, he will observe at the same time that this origin comes not from the front of these ventricles but opposite the middle of their length. From this it quickly becomes obvious how scarcely diligent Galen was in the anatomy of these organs when he imagines they are perforated and teaches that the cerebral ventricles end in their cavity to become like canals in which cerebral phlegm is taken to the area of the skull where these organs end and the hard membrane of the brain appears pervious like a sieve. These are things imagined by Galen, for as he inaccurately reported the beginning of those organs, so too he falsely described them as perforated and hollow no less than canals and ducts for phlegm are to be considered.

129   In fact, if we carefully study the actual straining-out of phlegm (as it is described by me) when a person is healthy in this part and is

not complaining of any symptom in its excretion or retention, it will be perfectly clear that phlegm is never borne anteriorly through these organs out of its own course without a symptom, any more than it flows posteriorly along the dorsal medulla and hence along the nerves into various limbs.

I must blame my own negligence that I do not find a foramen in the optic nerves; perhaps they have a foramen constructed differently from those of other nerves which can be understood by dissection. Whenever I perform a dissection of live animals or inspect the nerves of a human head that is still hot after an execution, or warm up in hot water the heads of persons who have been dead for some time, I never find the appearance of those foramina. In pigs, however, I find a difference because in them the optic nerve is transversely divided, and when boiled is constructed as if with many strands like the nerves that occur in the arms and legs. The optic nerve of man, however, is made of a continuous substance, uniformly dense everywhere, and in my opinion is not pierced by an obvious foramen. I think it is due to a faulty copy that in the book *De nervorum dissectione* it is written that the second pair of nerves is harder than the third. I am therefore unwilling to fault Galen in this case because I am otherwise nowhere able to surmise in the absence of copies where an error has been made.

Of the two branches running from the fifth pair of nerves when it is still in the cavity of the organ of hearing, toward the temporal muscle and branches of the third pair, mention occurs in Galen of only one. Likewise, he also missed the root of the nerve to the interior, which does not originate from the side with the fifth pair in the same way as what we call the lesser root of the third pair is brought forth next to the principal ✳ nerve (but the outer one) of the third pair. Because of this I would not add the root that I discovered to the fifth pair lest I disturb other anatomists' system of numbering for its sake.[172] This root

130

---

[172] Cf. *Fabrica* IV ch. 2: "Close to the root of the fifth pair, careful dissection has taught me that another pair [*n. abducens*] originates, unknown to all who make a study of

travels forward beneath the base of the cerebrum, passing through the hard cerebral membrane and claiming for itself a special foramen in the skull; it is then presented to the temporal muscle and principally to the muscle [*m. pterygoideus medialis*] concealed in the mouth.

If after closely examining the complete construction of the cerebellum we believe the beginning of the dorsal medulla comes from the cerebrum and that the cerebellum clearly joins it only by a single conspicuous connection at either side of the ventricle common to the cerebellum and the dorsal medulla, we cannot agree with Galen, who provides to the cerebellum so many nerve origins that must be assigned to that interval of the dorsal medulla which extends from the base of the cerebrum to the area where the occipital bone is articulated to the first vertebra.

Descriptions of the parts that are contained in the peritoneum, which are not entirely true

Above (so I may write something about things contained in the peritoneum), when I was setting forth examples of untrue descriptions in the muscles and mention was made of foramina in the transverse septum, I was concluding from a true description of the position of the stomach and the meeting place of its upper orifice with the esophagus, that the esophagus does not use the same foramen as the great artery which must run along the lumbar vertebrae. At present I must not fail to say that Galen wrote that the esophagus is not only conjoined with the descending great artery at the place where also, in his opinion, this foramen of the septum stands, above the eleventh thoracic vertebra; in addition, that the esophagus is attached to the artery as far as the twelfth vertebra, no differently than if the esophagus were first raised from the artery and the vertebrae of the back beneath the septum and went to the stomach by passing from right to left. Because

---

dissection. I shall not, however, depart from the old numbering of cerebral nerves" (p. 422, misnumbered 322).

the entire stomach is placed somewhat higher here than one would consider unreasonable, would the esophagus turn upward again, to be connected with the higher orifice of the stomach? ⋇

Therefore reason argues no less emphatically here than dissection that Galen accurately described neither the course of the esophagus nor its position. So far as concerns his description of the site of the whole stomach, Galen determined that it stands in the middle of the body so that he could represent all dimensions according to the mob of anatomists who are too casually trained in the proportion of man. But Galen also did not want to put that so glibly as not to state that the greater

131

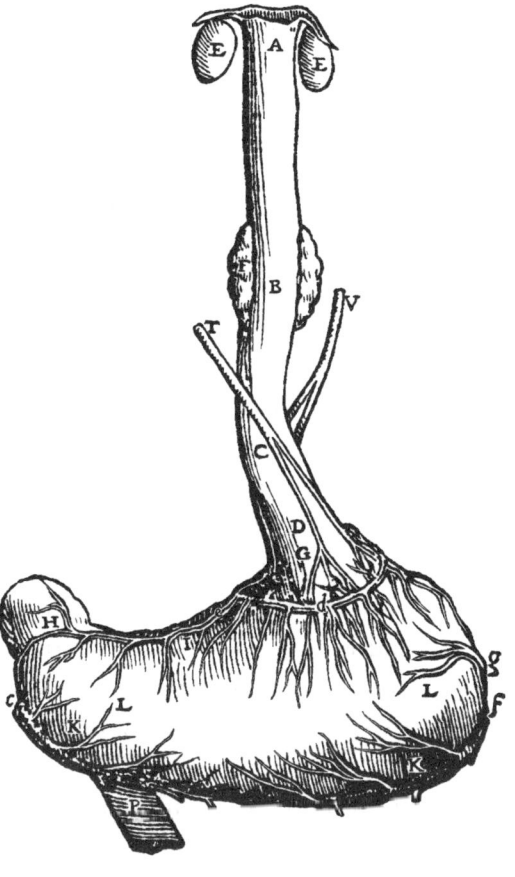

Anterior view of the stomach as shown in Fabrica V fig. 14. A, B, C, and D mark the course of the esophagus, the slight bend of which at C places the bulk of the stomach toward the left side of the body. H locates the lower orifice or pylorus, which Vesalius emphasizes does not emerge from the bottom of the stomach.

bulk of the stomach is placed on the right side of the body, as do all the compilers of anatomy who follow him, adding that this was rightly done so the stomach would be conveniently placed beneath the liver and kept warm by it, gaining from it an improved power of digestion. Surely, if we have examined the human body not in our dreams and imaginations but by careful dissection and inspection, we shall find the much greater part of the stomach in the left side of the body than in the right. For if we measure the width of the stomach from left to right we will note that as much as two thirds of it occupies the left side. This is quite opportune in respect to position, because the spleen lies beneath the back of the stomach low on the left side and with its point inclines only a little forward down to the left side of the stomach, and easily allows the left side of the stomach to be coterminous to the transverse septum over a large area. On the other hand, the quite considerable thickness of the liver prevents the right part of the stomach from coming into contact with the septum either on its side or its posterior surface or to a large extent its anterior, and it easily ensures that the right side of the body does not provide space for the stomach as the left side does.

Again, if we consider the true shape of the stomach and observe contrary to others' descriptions that it is not equally large along its entire width which I was measuring a little earlier from left to right, * we will see that because of its shape the stomach fills the left side of the body more. This is because the stomach is quite large and swollen on its left side, allowing for all differences of position, and the more it is moved to the right the more it is forced into a narrow space, and is seen much more constricted there and narrower. If one drew circles around the two sides, the one around the right side would look twice as small as the one making a circuit of the left.

I do not know whether in describing the shape and position of the stomach I should not have Galen consistently in the number of those who set the lower orifice of the stomach in its lower side: I have to this day found no one who rejects that opinion. I also say that nothing has been observed about the site of the stomach's orifices except that they are not opposite each other, or they are not both fashioned in the left side of the stomach; though in fact the orifice that we call the lower, the beginning of the intestines, is not at the bottom of the right side but at its highest point; and though the lower orifice is brought out of the stomach like an intestine, it rises a little before it curves along the back of the stomach. This must be examined without haste, in view of a great many problems about digestion in the stomach, the order of foods, and vomit, as these have now begun to be brought into dispute.

To this controversy belongs the opinion by which the fundus of the stomach[173] is considered fleshy, just as if its lower part along the width of the body were quite fleshy and thick, and altogether different from its upper region. I am able to imagine no difference, unless I were to decide that the area closest to the upper orifice is quite fleshy, not only because of a large number of nerves but also because of the quantity of veins and arteries occurring there, and because it is coterminous to the esophagus, which is otherwise more fleshy than the entire stomach. However, * neither color nor substance suggest that anything more fleshy be credited to the fundus.

I am quite surprised that when Galen examined the glandular body [*caput pancreatis*] extended toward the duodenum and wrote incorrectly that it was responsible for closing the lower orifice of the stomach, he did not think about the substance of the stomach in its orifices, where it is much thicker than in the rest of the stomach cavity, as if showing one circle protruding inward in each of its orifices which is seen to shut off the stomach from the esophagus and the intestines, bringing no slight toughness to the stomach and making it resistant to injuries if some rough, larger body should ever be brought to its openings.

If any vein offshoots are presented to the stomach from the vena cava, as Galen asserts in *De anatomicis administrationibus* that some are taken from there where it lies beneath the stomach, I gladly admit my ignorance, as I have so far discovered none. I also believe the reason is that no vein is distributed to the stomach except offshoots of the portal

133

---

173 The fundus gastricus, above and to the left of the entrance of the esophagus and thus its highest point.

vein. I am also such a newcomer to thought that I have never to this day believed the popular notion anything but pure fable that originated somewhere or another, according to which it is claimed that the esophagus is narrower where it connects to the stomach than in the rest of its course through the neck or the thorax because of the black, melancholic blood or juice which is belched up here from the spleen, and that the upper mouth of the stomach collects it and cuts it off.[174]

However, that has more to do with the use of the parts than the reason for their form. An example is the location of the omentum, which I find explained the same way by everyone:[175] they write that the front of the stomach is covered by it so that if cut away from the stomach and the intestines it would result in their weakness. But if trust is to be placed in dissection, the upper membrane of the omentum is attached like a suture to the lowest part and bottom of the stomach, * or rather it takes its beginning from there, nowhere lying upon the front of the stomach.

134

A similar case is the account given by those people of the colon: they say it covers the front of the stomach, writing that the colon rides upon the stomach, when in fact the colon is extended only along the lower region of the stomach, nowhere rising to the front higher than the stomach. Just as Galen wrote that some offshoots are presented to

---

[174] Having cut the ground from under the ancient doctrine presuming that a mythical black bile originating from the spleen is transported into the stomach, Vesalius does not pause to reflect upon its impact on humoral theories of character.

[175] See for example Alessandro Achillini's *Annotationes Anatomicae* (1520): "The heart is above it with the diaphragm between; below are the mesentery and the intestines, on the right the liver, on the left the spleen, in front the omentum." (tr. Lind 1975, 46); Alessandro Benedetti's *Anatomice* (ca. 1497) ch. 10: "The omentum protects [the stomach] in the anterior region." (tr. Lind 1975, 93); and Niccolò Massa's *Liber introductorius* (1536) ch. 14: "Around the middle of the stomach, going lower, is the origin of the omentum or zirbus." (tr. Lind 1975, 193). This idea can be traced to Aristotle *Historia Animalium* 495b29: "The omentum is attached to the middle of the stomach" (Loeb tr. by A. L. Peck). Vesalius took issue with these views at the end of *Fabrica* V ch. 3 in the section titled "No part of the omentum covers the front of the stomach" (p. 494).

the stomach from the vena cava, he also testified that certain offshoots are propagated from the vena cava into the omentum, in addition to the branches from the vena cava that implicate it in numerous ways; he likewise put it about that other veins are provided by the vena cava to the mesentery and intestines. This opinion occurs in the sixth book of *De anatomicis administrationibus*, where it is also read that veins reach into the mesentery that did not end at the liver. On the other hand, he argues that no vein unconnected to the liver is found in the body (however that may be true), disagreeing with Aristotle *De venarum ortu*. Furthermore, I believe that by that limit he distinguished offshoots coming from the vena cava from those which are implanted from the portal vein into the intestines and bring to the liver the juice which is the material of blood. However, I have seen no twig distributed from the vena cava into the omentum, the mesentery, or the intestines; I regularly advise students of this fact when after the stomach, omentum, mesentery, and intestines have been removed from the body during dissection, the stem of the vena cava passing downward lies in full view free of membranes and swollen with blood, and shows no offshoot cut from it which could have been presented to the organs just mentioned. To make this clearer, I would sever some little branch taken to the renal membranes or the seminal vein at its origin so that the immediate outflow of blood would prove that I had not carelessly cut any offshoot. I would not take sole credit for this but would earnestly entreat those present that if they ever discovered such veins, ✳ they should also not hide them from me, so that some day they should repay me, who have learned such different things but have been taught by no one, and so I could share the discovery among them without envy.

135

I have no doubt that Galen believed the liver is divided into lobes; it is a fact that he was responsible for its division into five lobes, which many under his influence identified with specific names, not as far as I know from surviving books of his. However that may be, Galen's account of the liver in man is not lacking in small errors (if only on account of its division). In the hollow of the liver, besides the small nerve derived from those distributed to the upper mouth of the stomach, I observe another[176] neglected by Galen which runs beside the artery presented to the liver and originates from the nerve of the sixth pair[177] of cerebral nerves running along the beginnings of the right ribs.

But if I were to make an account of the offshoots of small nerves, veins, and arteries, I should be overly occupied with an enumeration of my observations; what I have written here, which is beginning to be tiresome in its great occupations, would be drawn out beyond my intentions, as has already happened. I cannot, however, omit the vein which a cohort of anatomists starting with Galen contends somehow or another extends like a passage from the spleen to the upper orifice of the stomach. By this vein, they argue, melancholic juice that is excremental to the spleen's nutriment is delivered from the spleen to the mouth of the stomach, performing, they believe, a great function in augmenting the appetitive force of the stomach and restraining its strength. They say other things gathered mainly from Galen's account in which he praises Nature's intelligent design in draining the gall bladder not into the stomach but into the duodenum. I should like it to be known to those who believe I have missed the vein from the spleen that my diligence consists in its investigation, as I have looked for this vein not only in animals but also in man in both private and public dissections before I published my book *De humani corporis fabrica*, and after its publication at Padua, Bologna, and Pisa. Later, I took no small pains to investigate that vessel especially because of a certain dabbler[178] who after learning some practical anatomy with

136

---

[176]   *Truncus sympatheticus*, believed by Vesalius to be a continuation or branch of *n. vagus*.

[177]   Vesalius' sixth pair of cerebral or cranial nerves includes the modern *n. glossopharyngeus* (IX), *n. vagus* (X), and *n. accessorius* (XI); like Galen, he believed that all three nerves were one large "cerebral" nerve, covered by a dural sheath and traversing the jugular foramen.

[178]   The unnamed dabbler or smatterer (*sciolus*) could be Vesalius' onetime student Realdo Colombo, who succeeded Vesalius in the chair of surgery and anatomy at

my help (as he is illiterate), and having heard more than once in the medical schools that I could not find this passage, or what the mob of anatomists imagine is a vein, even though I had placed the stomach and spleen clearly before his and others' eyes, he cut apart some body when I was absent from Padua and boasted that he had discovered an unknown vein. He believed that after the publication of my book I would not return to Italy and would not compare what I had written with bodies in public dissections either at Padua or at Pisa (as you know).

When therefore I came to the spleen in the order of dissection,[179] and the bodies of a woman and a man were available at Padua, on the day before I had decided to move the stomach I earnestly asked everybody also to direct their eyes to a careful inspection; I asked them to bring along those who had laughed that this vein or passage, as they call it, was unknown to me, and I too would learn something from them. But as that dabbler, who is otherwise no careless viewer, was never absent, he was also mindful not to be present at the dissection when these parts or the muscles of the eye were being examined. I was therefore compelled to the same conclusions as before when we observed exactly the same things as previously. The spleen is connected to the stomach by the omentum, which is attached to the lower spleen by a membrane with vessels and constitutes a wrapping for it. Next, an upper membrane is common to the spleen and the stomach, delivering veins and arteries which take their beginning from vessels going to the spleen and are to be inserted next to the

---

Padua. During a brief return to Padua in December 1543 Vesalius heard himself criticized by his successor, "so that a generation later Vesalius was still referring to Colombo in the bitterest terms." (O'Malley 1964, 110). See Moes and O'Malley 1960, 508–28.

[179] Though Vesalius would note in the 1555 *Fabrica* Vesalius that "different orders of dissection are applied to different objectives" (Bk. II ch. 6 *ad init.*), anatomists regularly dissected soft tissues that would decay quickly before proceeding to harder tissues such as muscles and bones. Beyond that, the order of dissection was calculated to avoid destroying parts that the dissector wished to reserve for later investigation.

spleen; in humans these make up a large number. None of these proceeds from the body of the spleen, but as I just said ✳ originate from the vessels that are to be inserted into the spleen.

Just as the attachment of the spleen is made to the posterior and inferior region of the left side of the stomach, so also not only is an abundance of vessels more conspicuous there but also a larger vein[180] occurs there finally which in humans surrounds the left side of the fundus of the stomach supported by the upper membrane of the omentum. The other veins are much thinner and do not run out very far onto the stomach. I have never seen one higher than the rest extending in such a way that I could claim the upper orifice of the stomach is interwoven with it. We know very well that this orifice takes in veins from the offshoot which ascends from the middle of the trunk of the portal vein heading to the spleen along the back of the stomach, and it supplies the upper mouth of the stomach with many twigs; in dogs especially it surrounds the mouth of the stomach in a most elegant way, like a crown. Again, the highest of the veins passing to the stomach, of those that are closest to the spleen, nowhere resembles in size or length the vein that I was saying runs along the fundus of the stomach. Moreover, throughout its progress it has a companion artery, and it does not present itself in an array or mode of distribution different from the other veins implicating the stomach; it also is not filled with a peculiar blood which we could say is either thicker or darker. It also does not extend to the cavity of the stomach with its own mouths, as do the vessels of the gall bladder going to the duodenum.[181]

But to put it once and for all, it does not differ from all the other veins except that it is surpassed by many veins in thinness and brevity of passage. My opinion therefore remains the same as before, and it has been necessary for those to maintain the same views who were present

---

[180]   *V. gastrica sinistra*, D in the picture of the stomach reproduced on facing page.

[181]   On Vesalius' demolition of the anatomy on which the humoral theory of melancholia depended, see n. 174 above.

when I dissected after the publication of my book. For this and other reasons I am unable to conceive that the highest vein of those ⁎ that 138 enter the left side of the stomach (not to mention its companion artery) separately emit this impure blood and melancholic juice into the mouth of the stomach for the sake of such important functions, although at present we are not dealing with the function of the parts.

In the construction of the kidneys I find Galen's account incomplete. We have little more from his description than the report that veins and arteries pass through the body of the kidney, and that urine with a portion of bile which is present in the blood is transmitted through the dense, hard substance of the kidneys while blood is retained, and flows down into the urinary passage.

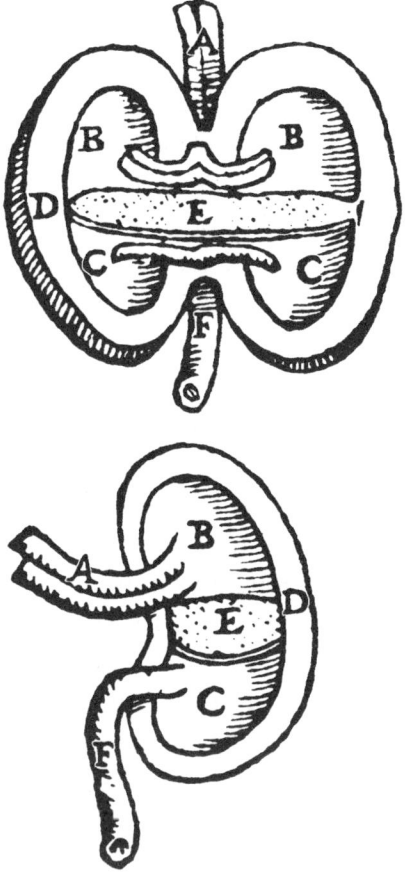

These imaginary sketches from Fabrica V chapter 10 (p. 515) show the conception that Vesalius rejects of a "completely trumped-up strainer or pervious, sievelike membrane [E] of the kidney, allowing urine together with bile to pass from the cavity marked B into the cavity labeled C." The ureter carrying urine down to the bladder is marked F.

If one considers what I have said about the construction of the kidneys, he will surely not take the flesh of the kidney to be like a sieve, as Galen is seen to reckon it, or as most people imagine a membrane in the kidneys, so easily perforated that it lets urine and bile through but retains blood as being thicker. However, I shall not repeat here the construction of the kidneys as I found it, as it is clear from the tenth chapter of the fifth Book of my work *De humani corporis fabrica* that it is the kind of thing that cannot be grasped in a few words (which is all I have available here).

Galen determined an origin for seminal arteries [*aa. testiculares*] that was the same as the veins, although we generally observe that each artery is brought forth from the front of the stem of the great artery lower down than the beginning of the right vein from the vena cava, while the left vein gets its beginning from the vein that is presented to the left kidney. But the intelligence of Nature that appears in the course of the right artery was not to be passed over in silence: she did not place the seminal artery beneath the vena cava in the same way we know the stem of the great artery is laid there beneath its left side, but she wished the seminal artery to pass over the vena cava elegantly, in a transverse direction. *

139   In the construction of the testicles, I wish Galen had explained to us what the distribution of the seminal vein and artery is like in the body of the testicle. We would have learned, thanks to the peculiar substance of the testicles, that semen is prepared from material brought to it in the same way in which it is known the substance of the liver makes blood. Likewise, it would not have been unknown to us how then semen is taken by the passage that delivers it from the body of the testicle and its vessels.

I shall say nothing about the uterus because I should be too lengthy, since I understand that Galen chiefly described the bovine uterus and not the human. Several people, however, have written me about my account of the uterus, especially my discussion therein of the menstrual purgations; mention was made of your letters, my Joachim, noting that nothing had been said about the hymen. I said nothing purposely, because I knew nothing with certainty: I had never dissected a virgin, except one girl perhaps six years old who had died of consumption. I had gotten her for preparation of a skeleton with the help of a student at Padua who had stolen her from a grave.

While I was cutting everything away from her bones, at it happens, and not taking the time to inspect any part, I did however dissect the uterus close to the hymen. Though I found it as I have recently seen it after the publication of my book, I did not dare to say anything about it because I perceived that animals do not have a hymen. So I never take a position on the basis of one dissection or another, and whatever harsh remark Terence makes about first coition[182] I had consigned to the sort of connection in which we know muscles lying upon each other are put together.

I will pass over the concourse of veins which the Arabs enumerate which is called a cento or patchwork by their translators. But I should have said more about the opinion * that the hymen in small girls is not entirely hidden when they spread their legs to urinate. If it had been an established fact that this membrane or fleshy septum had been called the hymenaeal, I have no doubt that this would have been agreeable to the one who named it.

When I was about to teach anatomy at Pisa there was a shortage of bones, and I believed an anatomy should be performed in exactly the same order as I had described in my book *De humani corporis fabrica* at the new inauguration of the great university;[183] the anatomy should be compared by the students with what I had written. At the command of the illustrious Duke of Tuscany Cosimo de' Medici (as he was given by the Gods especially for the advancement of studies, and he left nothing lacking which could accommodate the students of his university), the cadaver of a nun from some hospital had been sent by swift boat for the preparation of a skeleton.

---

[182] Probably a reference to *The Eunuch* of Terence (161 BCE) in which a 16-year-old boy named Chaerea disguises himself as a eunuch to rape the virgin Pamphila, also 16. The rape is not described except that he tore her clothing and pulled her hair, and Pamphila is too traumatized afterward to be able to answer questions about the assault.

[183] The ceremonial re-opening of the Studium at Pisa was enacted on November 1, 1543.

There was also at the time among some of the students a supply of keys to the rare and elegant cemetery of Sanctus Pisanus so that they could search among the monuments constructed in a kind of sanctuary if there was anything useful to them for the inspection of bones. The best suited are the ones which are placed transversely among the monuments of this cemetery and therefore admit rain and air. Because some are in monuments beneath the surrounding roof, the bones are less useful for study because of decay and the adherence of ligaments. In one of these, whose epitaph appeared recently made, a hunch-backed girl was laid who had passed her seventeenth year and had died, so far as I could guess, from impeded breathing due to the bad formation of her bones.

It could easily be determined that the nun had died from pain in the side,[184] an inflammation that occupied almost all the entire left side of the membrane enclosing the ribs, but especially at the roots of the ribs.

[141] Similarly, in the same year an elegant prostitute was removed by students from a monument next to the church of St. Anthony and brought to a public anatomy in Padua; she had died on the third day of an inflammation that followed the entry of the unpaired [azygos] vein and its offshoots and had occupied the entire back of the thorax, providing us an outstanding specimen for recognizing the character of the lateral disease. The rest of her body was only slightly emaciated and was therefore perfect for dissection, which was the last one I performed at Padua.[185]

So when the cadavers at Pisa were freed of their flesh, the nun and the girl whose bones were for preparation of a skeleton, I

---

[184] Vesalius calls the disease *lateralis morbus* or *dolor lateralis*. This "pain-in-the-side" disease includes pleurisy and related pulmonary disorders; its treatment in Vesalius' time consisted chiefly in venesection. For a discussion of what the *dolor lateralis* was, see Saunders & O'Malley 1947, 8, and Smith 1990.

[185] *Fabrica* V ch. 15 (p. 539) mentions two other female cadavers acquired after this one, but not necessarily used in public dissections. One of them, he says, was the model for figs. 24 and 27 of the *Fabrica*.

inspected the girl's uterus with a few students who were present at the time because I inferred she was a virgin, particularly because no one had solicited her favors. I did find a hymen in her, as I did in the nun who was perhaps thirty-six years old, no less with the damaged testimony of her by now constantly stressed parts.

Moreover, when we were returning from our expedition to France[186] I was asked by the doctor of Countess Egmondana to attend the dissection of a noble girl eighteen years of age whose uncle suspected she was killed by poison; she had been long disfigured by a pallid complexion and drew breath with difficulty (though otherwise quite elegant in appearance). Because an extremely unskilled barber was performing her dissection, I was unable to keep my hands from the task. At other times I had never watched another dissector since the two inexperienced dissectors whom I first saw at Paris, and three days before that at the university in free dissections. So from the narrowness of the thorax and of the corset which the girl used in order to be notable for her slender, slim, long torso, I believe she deteriorated because of compression of the breast around the abdomen and lung. ✳

Since damage to the lung and a marked compression of the viscera in the abdomen had revealed the cause of death and nothing had come to my attention by which I could distinguish strangulation of the uterus except by swollen ovaries, I dissected the girl's uterus with the doctor for the sake of the hymen when the attendant domestics had left with a few spectators to make quick disposal of her foundation garments. However, the hymen was not entirely obvious to me, though it was not so hidden as I usually consider it in women under sixteen[187] where it is at other times situated, just as if the girl had broken her hymen with her fingers or in some other wanton

---

[186] Following the Treaty of Crépy of 18 September 1544. *Comitissa Egmondana* is probably Anna van Egmont (1533–1588), only daughter of Maximiliaan van Egmond. In 1551 she would marry William the Silent.

[187] We have emended *vix sedem* here to *vix sedecim*.

way without using a man, or according to a remedy from Rhazes for strangulation of the uterus.[188]

In the neck [vagina] of a girl's uterus, therefore, soon after the beginning of the neck [urethra] of the bladder into the upper part of the female fold[189] there is a certain transverse septum consisting of a fleshy, skin-like substance very like the one of which we see a water-lily is formed. This septum is attached at its thickness to the sides of the neck of the uterus, a little thicker at its connection than over the rest of its surface, though not much thicker than the rest of the membrane. In the middle of the septum there is an opening cut like the pudendum in a long slit so it will not retain the menstrual purgations in virgins; there is no need to imagine any veins below this septum or hymen from which menstrual blood would drain in virgins or which should be distinguished from the veins that otherwise discharge the menses of women, meaning the difference between pregnant and non-pregnant women.

When Galen described the wrappings of the fetus, he said they resemble the wrappings of dogs and pigs; I have easily shown elsewhere how well his descriptions describe dogs. I would now do likewise if I were as practiced in examining the wrappings of the human fetus * as I am those of dogs. Since the publication of my book I have still had no opportunity to dissect a human fetus still preserved in utero. When the opportunity arose once or twice before publication, I was so ignorant of these matters, like other doctors in attendance,

143

---

[188] Most likely a reference to the ninth book of Rhazes' *Liber ad Almansorem*, which Vesalius had published in paraphrase in 1537. It covered pathology and therapy of small ailments of the body, and had been most recently published in Latin translation at Basel in 1544. As early as the Hippocratic *Peri Parthenión*, a variety of physical and mental ills was attributed to strangulation of the uterus resulting from blockage of the menses in virgin girls.

[189] Here in Vesalius' Latin the vagina is the *cervix*, the urethra is a *collum*, and the "female fold" is *muliebris sinus*, probably the vestibule of the vagina inside the labia. Vesalius seems here to have revised his understanding of the urethral opening, which he had placed inside the vagina in *Fabrica* Bk.V fig. 27. But see n. 209 below. The word for water-lily is *nymphaea*.

and the work had to be so hurried, that there was no opportunity to observe the differences between dog and woman.

There is, in fact, some difference in the fleshy substance that is like a fascia in dogs and which for that reason I was sometimes reckoning as the outermost wrapping. In women, as I sometimes learned when called to women in difficult childbirths, this fleshy mass, which is quite similar in substance and makeup to the spleen, nowhere surrounds the entire fetus, though it is continuous like the body in dogs but not spread about as it is in cows, cervids, and other horned animals. Upon these there are nearly uncountable bits of this fleshy substance distributed through the membrane surrounding the entire fetus in about the same way as the dark spots appear on the back of a leopard.

The illustration on the left was added to the 1555 edition of Fabrica V (fig. 32) to show the dark spots (cotyledons or acetabula, marked B) on the placenta surrounding a calf's fetus. In the 1543 edition, Vesalius had illustrated the canine annular placenta (fig. 30, right) which is no 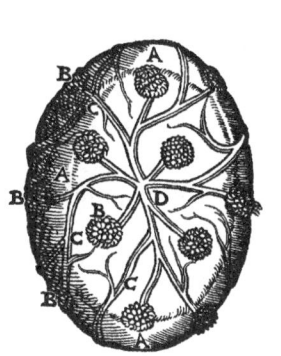 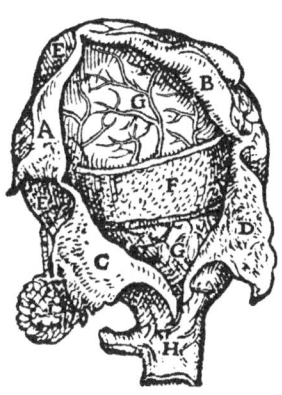 *more than a horizontal ring (F) around the second wrapping or allantois (G, G), which is vestigial in humans. He had still never examined a human fetus in utero.*

However that may be, it is clear enough that Galen's descriptions of the wrappings fall very short of the truth and a complete account.

Several untrue descriptions gathered from the parts contained in the thorax

In his account of the wrapping of the heart (if I may add something about the things that are placed in the thorax), Galen stated wrongly that it originated from the base of the heart, though the wrapping is no less distant from there than from the point and the remaining surface of the heart. There is also an ample interval between the base of the heart and the attachment of the heart's wrapping where it is joined to the vessels proceeding from its base. The vessels themselves do not have this tunic or wrapping in the interval between because they acquire it from the adjacent membranes: the vein gets the second wrapping and the artery the third. ✳

144   But it is still more of a surprise that Galen examined the construction of the heart so hastily that he wrote that the orifice of the venous artery [*vena pulmonalis*] is smaller than the orifice of the great artery [*aorta*], citing reasons why he should reach this conclusion no less effective than actual dissection, which handsomely shows the opposite. The orifice of the venous artery is much larger and wider than that of the great artery, as Nature did not build it contrary to reason and uniquely good design. But Galen not only argued that the orifice of the great artery is larger than that of the venous artery, he also decided that it is much greater than all the orifices leading into the heart, comparing it also to the orifice of the vena cava, though in fact without any falsehood it is seen that the orifice of the vena cava is even twice as large as the orifice of the great artery.

It is impossible for someone to place no importance on these descriptions by Galen that are utterly inconsistent with the truth. Or even when one has learned of his carelessness in dissections, he may blame his editions of Galen. For as a result of his false statements about these orifices one may gather as many arguments against Aristotle and sometimes Erasistratus, and again in praise of Nature. Though these are powerful statements, it is finally well known to a person who carefully inspects the fabric of the heart that he admits the greater propositions of Galen; then he accepts in a lesser way outside the truth of the matter something other than he believes, afterwards determining a true conclusion at odds with

Galen's opinion. Since such opinions have come to my attention in Galen which contradict the opinions of the Ancients, and as I am compelled to admit that Aristotle is much more deserving than Galen in studies that are common to both – though I swear by the words of Galen[190] – should it seem strange to someone if something irreverent ✳ escaped from me against Galen? Careful as I have been at all times to prevent this from happening, as long as I live I shall try to avoid having anyone say that about me truthfully. I am ever so annoyed by those who are in a hurry to purge Galen of false reasoning. You know with how inconstant a mind, and one not entirely dedicated to Galen, those frivolous objections which are sometimes not without malice are able to provide the occasion for someone to become zealous to accumulate errors in Galen, collect them into a single volume, and publish them. I shall never attempt this, since I respect Galen more than any mortal. Although I show what he did not think about, this is only so that when I explain true anatomy no one will think because of his authority that I am pointing falsehoods out to people.

    This happens in the description of fibers that are brought out of the membranes controlling the orifice of the vena cava which I believe are attached along the sides of the right ventricle of the heart to the substance everywhere at its point; Galen on the other hand said (though falsely) that they are attached only to the ventricular septum of the heart.

    Similarly, in the small glands of the throat someone would believe that I was imagining a third type of glands if I had not added to my account that Galen had missed the small glands resting at the

145

---

[190] An ironic tag from Horace's *nullius addictus iurare in verba magistri* (*Epist.* 1.1.14, "given to swearing by the words of no master"), from which would later come the motto of the Royal Society of London, *Nullius in verba*. Vesalius had begun using this tag (without irony, labeling himself a habitual skeptic) in *Fabrica* V ch. 13, and he repeated it three more times in Books V and VI.

root of the larynx where its second [cricoid] cartilage was attached to the remaining stem of the trachea; these glands have a large number of their own glandules, noteworthy and unlike any other type of glandule.

False descriptions among the parts that are surrounded by the skull

Likewise when I survey the process of the hard membrane that comes between the right and left parts of the cerebrum and also the one that separates the cerebrum and cerebellum, I deny that they are doubled (because in fact they are single and correspond in substance to the sides of the hard membrane). I might perhaps be blamed, because ✳

146 Galen imagines the hard membrane is doubled at that point, being confused by the sinuses that stand in the area of those processes where they are joined to the skull. Again, when I pursue the study of the cerebrum and say that it is continuous I have to warn

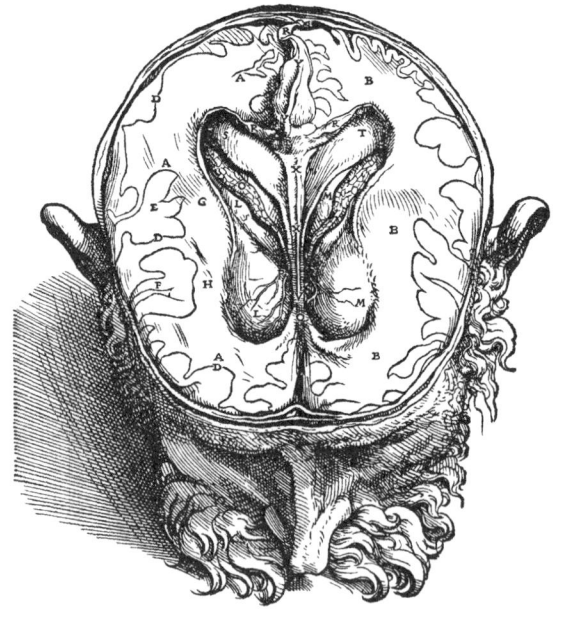

One of several cross sections of the cerebrum in the Fabrica (Bk. 7 fig. 5, p. 609) showing parts that are common to the left and right sides. The corpus callosum (R, R, R) has been cut away from the front of the brain and folded back, exposing the upper surface of the fornix (S, T, V). The colliculi are still out of sight beneath the septum pellucidum (X, X) that separates the two ventricles (L L, M M).

the reader that it is not so divided into a right side and a left that it is nowhere continuous.

That is because Galen's account, where he explains the reason for the brain's division, sounds no different than if the cerebrum were doubled like the two eyes and two ears that we know Nature gave us, though in fact the cerebrum is continuous at its base not only because of the body that we call the corpus callosum but also because of the one [*fornix*] that we compare to a tortoise shell or vault, the one we compare to buttocks [*colliculi inferiores*] and testicles [*colliculi superiores*], and finally the large portion of the cerebrum itself from which the beginning of the dorsal medulla issues.

In describing the beginning of the dorsal medulla, I could not fail to remark that Galen described it as if it did not hang down from the cerebrum and as if its entire substance were not continuous with the cerebrum. When Galen inquired into the reason for the construction of the cerebellum and determined that it was simple and single for the sake of the dorsal medulla (which needed to be single and not double like the cerebrum in its upper area), he set about his account just as if the cerebellum were not continuous with the cerebrum and also put forth a single dorsal medulla. It may be seen in my writings how foreign these ideas are to the true fabric of the body; one will immediately understand that they are false, as I just said, because Galen taught differently from me – though when admonished (if he were studious of the truth) he would moderate his opinion so long as he learned from dissection and examination of the body how the matter stood.

The situation is the same when I say ✳ the anterior part of the right and left ventricles is blunt and large; they are not confined to a narrow space there, and they end at the olfactory organs, which like canals deliver phlegm to the [ethmoid] bone of the head that is permeable like a sieve. Much of this, which Galen reported, is quite foreign to the true fabric of the cerebrum. Among the

147

errors, the least to be neglected is that when Galen pointed out that the ancient professors of dissection, when describing the processes of the thin cerebral membrane, stated that portions of this membrane were borne into three cerebral ventricles, he quickly imagined that these ventricles were lined and coated with the thin membrane, and deduced that the reason for this was the softness of the cerebrum, as if the ventricles would have collapsed unless they were supported by the thin membrane. This is the difference by which Galen distinguished the fourth ventricle (which we locate in the cerebellum) from the three preceding: that is, because it is harder than would require it to be lined and supported by a thin membrane. However, the previous ventricles are no less invested by a thin membrane than this fourth, and the fourth ventricle was said by the Ancients to differ elegantly from the previous ones because no process of the hard membrane presented itself therein. Therefore networks like a placenta are visible in the previous ventricles as well as vessels entering into the constitution of those networks.

Finally, I would like to avoid spending any more time on false descriptions, but I have not been able to pass over the errors of Galen which occur in great numbers in his description of a reticular plexus.[191] Of all the parts in the body there is none as talked-about by doctors and philosophers as this blessed and wondrous network. In its description I was unable to accept that the carotid artery where it is closest to the skull is brought whole into the head since it sends such a noteworthy portion backward ✳ which is drained into the sinus of the hard membrane [*sinus transversus dexter*] that I call the first. I will not now discuss the fact that Galen was altogether ignorant of the foramen [*canalis caroticus*] by which the greater part of the carotid artery is borne elsewhere inside the skull cavity. However it was that Galen traced its path in his account into the space of the

148

---

[191] See n. 141 above.

head or the skull, he attests that all of it is taken up into some net-work which he compares to nets lying upon each other, without noticing the large offshoot presented to the sides of the hard membrane or the noteworthy portion extended to the nasal cavity; he also does not notice the large twig which goes with the second pair of cerebral nerves to the eye socket and the eyes themselves. So it is that it should not come as a surprise if I do not agree everywhere with Galen's description, and many things appear in his books from which I cannot quote examples of true descriptions as I would like to clear myself of Sylvius' false charge and provide his disciples with material fit for sharpening their pens – if only they were willing to be occupied otherwise than in exalting Galen in their praises (which I do not think is my task), snarling in unison without recalling all the places where they think I have censured him, and chattering calumnies and insults. They should together consider whether they have taught others what they are saying, or have dissected with their own hands in public universities in the presence of the most learned men, not ones who are still untrained in medicine and students who are just beginning.

Some places where it is known that Galen was not altogether sound in assigning the functions and uses of the parts

At present it is claimed that I have falsely put it out that Galen made even the smallest mistakes in accounting for the function or use of the parts. I can refute this charge, so may the Gods love me, with a much larger number of reasons than was available for the two ideas of Galen previously mentioned. If time now permitted, * as it did when I was living in Academia, and I was dedicated to our studies and not to the labors of my craft, and I had the choice of sitting idle at home, I would scrutinize at much greater length than I did in other places that have

149

been criticized the uses and functions wrongly assigned by Galen and actions that have been incorrectly explained. However that may be, from a great heap and storeroom, so to speak, I am willing to add at least a few places so you may have an example from this portion of my reply to the letter of Sylvius.

In his account of the bones

I will pass over several foramina of the skull in assigning whose use Galen's mind was wandering no less than in his descriptions, as in the foramen[192] transmitting the second pair of cerebral nerves; in the foramina carved in the cribriform bone [*os ethmoidale, lamina et foramina cribrosa*], which we have written is located in the base of the frontal bone; in the foramen [*f. ovale*] transmitting the third and fourth pairs of cerebral nerves; in the foramina [*canales carotici*] made for offshoots of the carotid artery, which came at the end of what I was saying above; and in the use of the bone [*os sphenoidale*] resembling a wedge at its base, where he said wrongly it is pervious like a sponge[193] and thereafter imagined both actions and functions there without any logic (as has been shown elsewhere). I should explain motions of the head, unless you think something should be brought in first by way of general descriptions of the bones. An example is where Galen is summarizing the use of epiphyses and writes that they are given to bones to be covers for the cavities in which the marrow is held. He did not consider that in bones that have such a cavity Nature did not place an epiphysis over those cavities where the reason for their composition and the series of various processes and protuberances required

---

[192] *Fissura orbitalis superior.* In *Fabrica* I ch. 12 (p. 50) Vesalius had written that it is larger than the foramen for the optic nerve, disagreeing with what Galen had written in Book 9 of *De usu partium* (3.718.14).

[193] "Galen wrongly left it written that this bone is perforated like a sieve or sponge and transmits cerebral phlegm." *Fabrica* I ch. 6 (p. 32), citing Galen *De usu partium* Book 9; see 3.694.1ff: "The gland is succeeded by a bone like a colander, that terminates at the palate, and this is the route of the thick residues."

particular hardness, as happens in the lower part of the humerus and the upper region of the ulna.

Also, an epiphysis does not develop on bones in which those cavities ✳ are large and hollow without an interval of bones, unless the bone itself has already become softer than it is along its entire length, is beginning to have a substance identical to the epiphysis, and the cavity is hardened prior to the common material of the epiphyses. Moreover the scapula, which nowhere shows such a cavity, has a great abundance of epiphyses, just as do the bones attached to the sides of the sacrum; these bear an epiphysis covering the entire spine of the ilium. A large epiphysis is also given to the hip bone where it provides a beginning to so many muscles moving the tibia and the femur.

Again, how many epiphyses abound on the vertebrae? They do not, though, have a cavity in which marrow is placed by itself. I know that marrow, or at least a juice which we have to compare to marrow, is contained in all bones having a spongy substance, and in their epiphyses. In assigning this use of the epiphyses, Galen should have considered how in small children the epiphyses are mostly cartilaginous, and then how their substance differs in softness from the other part of the bone, so that he would not invent the idea that Nature constructed epiphyses or covers of cavities containing marrow.

So far as concerns motion of the head over the first vertebra, and of the first vertebra with the head over the second vertebra, and then of the head with the entire neck, I cannot sufficiently wonder why I should be condemned by anyone for not agreeing with Galen and for putting forward something completely at odds with what he wrote:[194] nobody's mind should be so imprisoned that he should not set aside some fondness and descend to my view, unless he is put off my ideas because they were proposed by a young man and he himself had

---

[194] On this subject see *Fabrica* I ch. 15, "Galen's opinion about the motions of the head" (pp. 63–5), where Vesalius mounts an extended critique of Galen's view.

read and corrected the books of Galen so many times and translated them * into another language that he was ashamed not to have previously observed these facts himself, particularly in the motion of the head, in whose description Galen requires a more learned and attentive listener than in any other matter, however serious and elegant. In addition, it must be admitted that so many uses of muscles and ligaments were badly explained by Galen, since he erred in describing the motions of the head.

I do not know whether Sylvius was influenced by the fact that while making this argument in my book I wrote that I had been left without the aid of a preceptor[195] – though perhaps he believes I learned anatomy from him, who still contends that Galen wrote nothing wrong. As long as I live I should note that Sylvius began in his fashion to read us Galen's work *On the Use of the Parts*. But when he came to the middle of the first book, and the anatomy, he declared that to be more difficult than we could follow as medical students, and therefore, he said, it would equally torment both himself and us. Then he began the fourth book as far as a portion of the tenth. After that, again leaving out the material in between as far as the fourteenth book, he also read the later books in such a way as to cover a book a day or in five or six days, never warning us that Galen had said something different elsewhere (as often happens), or showing that something written by Galen was contrary to how the matter stood, at the same time bringing into class parts of nothing except a dog. In dissecting these parts we students were so painstaking and imitated the best teacher, that he never once tested our diligence after his lectures. So it then happened that on another day we showed him the membranes controlling the orifice of the arterial vein [*truncus pulmonalis*] and the great artery [*aorta*] which on the previous day he had assured us he was unable to find. *

---

[195]   See *Fabrica* p. 63: "It should seem strange to no one that I too was very much in need of a preceptor's aid in this part."

Since Sylvius had skipped the places dealing with the verte- 152
brae along with many others in his so-called course of 1535, and had
never read us another anatomy
book (except *De motu musculo-*
rum, in which he also thought
Galen's opinion was consistently

The first three cervical vertebrae seen
from behind in fig. 11 of Fabrica I ch.
15 (p. 60). N is the articular surface
of the atlas or first vertebra, joining
the neck to the occipital condyle of the
head. Γ *identifies the dens or tooth of*
the axis or second vertebra, creating
the swivel on which the head rotates.

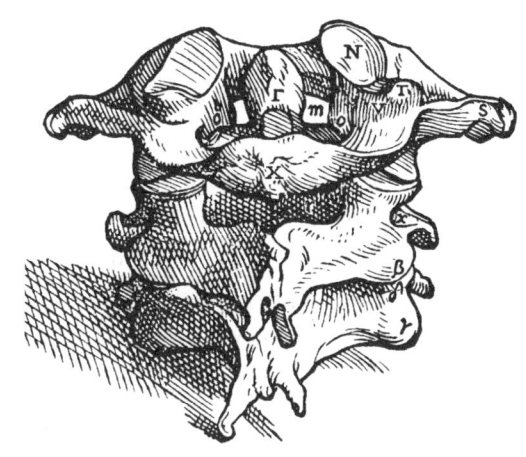

correct), it is no surprise if I write that I lacked the service of a preceptor and was especially watchful to be counted among those lis-
teners whom Galen did not repel from his books when explaining
motions of the head and who believes those movements are suitable
for understanding the mysteries of Nature.

So it came about that I
said the head is flexed in its own

This figure from p. 66 of the Fabrica
illustrates the ligaments that restrain
the dens of the axis within the hollow
of the atlas or first cervical vertebra (A,
B, C). The dens (H) is seen within
the cruciform ligament, whose supe-
rior band is marked I. The transverse
ligament of the atlas is marked K.

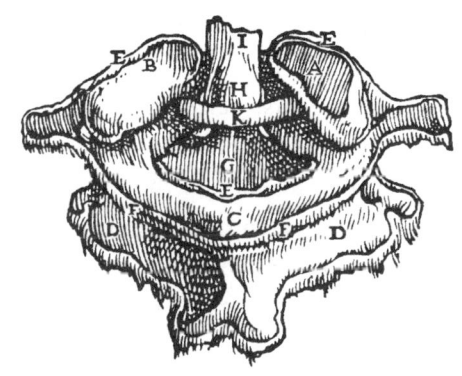

motion above the first vertebra and is moved upward or to the rear,
while Galen teaches that the head is moved to the side above the first

vertebra as if we were inclining it to the shoulders. He introduces nothing else explaining motions of the head, particularly in the place where he states that the condyle of the occipital bone is elevated from its depression in the place that stands directly opposite the side toward which the head is moved. Not only is Galen content to have described the motions badly, he also attributed such negligence to Nature that in the worthiest joint of the entire body she wished for the bones to draw apart and separate from each other in the motion – though in fact for the best of reasons we never see an empty space made between the joints in any motion.[196] It is therefore clear enough that the motion to the sides is attributed by Galen to the joint of the first vertebra with the second, though in fact the head is not moved to the side by its own motion but only indirectly with the entire neck, which is moved gradually in an arc to the side with all the vertebrae consenting to this motion.

Still less skillfully, Galen taught that the head is flexed forward and back with the first vertebra over the second, in the motion that I said is performed in the first. ⋇

153 Clear language in three different books attests what Galen's opinion was about that; a still worse mention of the use of a ligament states what prevents the dens of the second vertebra from crushing the dorsal medulla in anterior flexion of the head. The fact is that the ligament restrains the dens so it is sequestered in the hollow of the first vertebra in the way that an axle is known to be rotated upon by a sphere or wheel or some such thing. The head (whatever Galen teaches, or the disciples of Sylvius sharpen their pens and write while paying no attention to the nature of the dens) is rotated over the second vertebra and is in no way flexed or extended over it; Galen in the meantime mentions no joint on which the head is rotated.

---

[196] These criticisms of Galen are made at greater length in *Fabrica* I ch. 15 in the sections titled "Galen's opinion about the motions of the head" and "A different opinion from Galen's about the motions of the head." A later section in the same chapter, titled "The worthiest joint in the whole body," about the joint between

If anybody were to argue that Galen understood it as a motion of shaking the head and assert that this was his opinion, that the head is rotated over the first vertebra, such a person will soon see what an absurd suggestion this is even for the sake of a joint: for the two condyles of the occiput go into two deep sockets laterally opposed to each other so that the head can not be rotated over the first vertebra, any more than a compass whose two legs are fixed in a stake. It is as clear to me that I am saying here what fits the truth and that the structure of the bones is consistent with my views, as I strongly believe that as soon as this attack of Sylvius cools down he will doubtless agree with me – if only he would carefully display the bones to his disciples and inform them of contrary views so he could observe with them how Galen's badly described motions of the head confuse many things in his account. None of these prevents us from confirming that Galen was not altogether perfect in assigning the use of the parts.

My writing would be unnecessarily prolonged if I were to revisit everything here about the muscles ✳ and ligaments that necessarily follow a misunderstood motion. I would deserve thanks even for this observation were not Sylvius so agitated against me in his letter. When Galen mentions the process [*tuberculum anterius*] seen in the front of the first vertebra, shaped for the strength of that vertebra and the insertion of muscles, as we rightly stated, he should not have forgotten his doctrines there and assigned it the function of pushing the head upward when it is inclined forward.[197] The head cannot ever be inclined so far that it rests on that process in such a way that it would perform that function. Anyone by whom the way bones are put together is understood has no doubt that

---

the first vertebra and the occipital condyles, specifically faults a passage in the 12th book of Galen *De usu partium* (4.1 ff.).

[197] Galen *De anatomicis administrationibus*: "The anterior arch of the first vertebra prevents the head from slipping too far forward, fixing and raising the head just before it goes too far." chapter 8 of the fourth book, §461, tr. Singer 1956, 113. Vesalius rejected this view in *Fabrica* I ch. 15 (p. 64).

this is quite clumsy and in any case has nowhere been attempted by Nature.

Galen, who occupies himself so much in explaining why the two lower thoracic vertebrae lack transverse processes, should have noticed that his opinion is false unless he had explained by the same token that the two vertebrae resting upon the tenth also lack the same processes. But as those vertebrae have the processes, so too Galen's arguments in this place are, as I explained above, untrue. As in anatomy Galen often made contrary statements, so it should not be surprising if sometimes he teaches that the same part was fashioned for contrary functions. An example is found in the acute process of the ulna [*processus styloideus*], in whose description Galen has been shown to have strayed far from the human fabric. That process is not extended beyond the remaining series of rims of the depression to which the carpus is articulated and is not coated with smooth, slippery cartilage, nor does it enter the socket of the third carpal bone as it would in a joint.[198]

As Galen fell far short of the truth in describing this joint, so too in his book *De ossibus* he wrongly ＊ wrote that the hand is moved obliquely by the joint of that acute process with the wrist, meaning I suppose the motion by which we move the hand to the side. If that styloid process went higher and were articulated as it presents itself in sheep, or as Galen writes, it would act like a stake and prevent the hand from moving to the outside; so far is that from the motion that Nature made it control. In his book *De usu partium* he not only attributed that lateral motion to this acute process but he also made it responsible for pronation and supination. Because that agrees with the truth, it must be distinguished from the motion of a hand which in no sense is moved to a prone or supine position by its own peculiar

---

[198] This paragraph recalls a theme previously mentioned: see p. 104 above. Vesalius makes this argument more fully on p. 114 of the *Fabrica* (Bk. I ch. 24) where he details the cartilage separating the wrist from the forearm.

motion. So poorly did Galen follow the skill of Nature by which the wrist joint as much as possible related only to the radius, by means of which the hand fitly undergoes derivative motion together with the radius; the sharp styloid process would prevent such motion if it matched Galen's descriptions. Similarly, we see that sheep and calves lack this motion of the joint because of the partnership, so to speak, of the ulna with the radius. So I should deservedly be judged mad if I so openly defied my reason as to say with Sylvius that Galen was wrong in his description of no part, function, or operation.

Because Galen taught no difference in the joints of the fingers, it is also no surprise that he did not think about Nature's craft in the lateral motion of the fingers. Because it is now very certain that the longest foot did not fall to man's lot but rather the shortest, no one should be in doubt how many faults can be found in the third book of *De usu partium* if he wants to spend time on matters that are all too clear and at the same time frivolous. ✳

Several uses and functions not well assigned in Galen's account of the muscles and ligaments

Earlier, when we pointed out the unnecessary blending of ligament and tendon that Galen imagines in the muscle (reasoning sometimes from elsewhere, sometimes from muscles that nowhere have their fibers more separated and scattered from each other than in their origin and from muscles that are seen to be as large in their origin and mid course as in their insertion), it was also easy to notice that flesh is not given by Nature to the muscles chiefly to be a fulcrum and layer of some division created to contain the fibers; but if we duly consider the construction of a muscle we think of flesh as its principal substance just as we think of a particular substance not common to any other part in the lung, the liver, the spleen, the kidneys, testicles, brain, heart, and in all organs in charge of a peculiar function. The material of ligaments is given to a muscle for strength, so there would be something which when contracted by the gathering of flesh would pull upon that into which the muscle was to be inserted. The muscle has need for a nerve chiefly for animal force,[199] which it provides just as we notice venous blood and arterial material is delivered, if we are dedicated to a zeal for truth and do not set too much store by authors. Just as we see a set of nerves paired with veins and arteries, we also see that it is suitable for a muscle that animal force be imparted most of all to the place where the muscle needs to be gathered into itself and where it will act upon the part to be moved. I believe I have argued about these matters elsewhere, so I will not need to linger on them now.

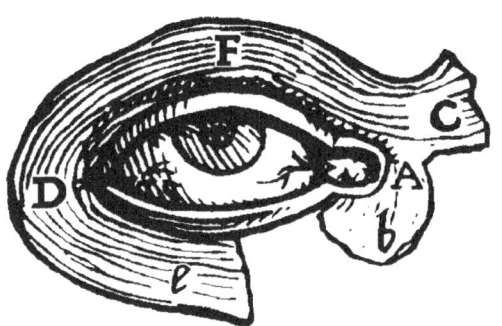

Vesalius' illustration of the eyelid muscles at the beginning of Fabrica II ch. 10. His first muscle (lig. palpebrale mediale) is labeled A, C, b; the second (m. orbicularis oculi) is marked D, e. The two meet at F. Vesalius added a short chapter to the 1555 edition (ch. 10) on the dissection of these muscles.

As Galen is inconsistent in counting the muscles of the eyelids, he also disagrees with himself about their function. In his book *De usu partium* he counts two muscles, one at each angle of the eye, of which he says the one that occupies the inner angle of the eye moves the eyelid upward; ✳ the one in the lesser or outer angle, he believes, causes downward movement of the eyelid. In his book *De locis affectis* he writes that those two muscles move the eyelid downward, and

157

---

[199] On this vaporous *animalis vis*, see n. 156 above. The existence of separate animal spirits was to be challenged in 1556 by Vesalius' contemporary and fellow neoteric Giovanni Argenterio (1513–1572), *De somno et vigilia* Book 2 chaps. 7–10. See Siraisi 1987, 339 and 1990, 177.

Vesalius' illustration of the eye muscles from Fabrica II ch. 11 (p. 239). The muscle labeled O, the retractor bulbi, is a bovine muscle not present in humans. The other muscles (H, I, K, L, M, N) have been pulled away from their usual position covering the retractor bulbi. B identifies the optic nerve.

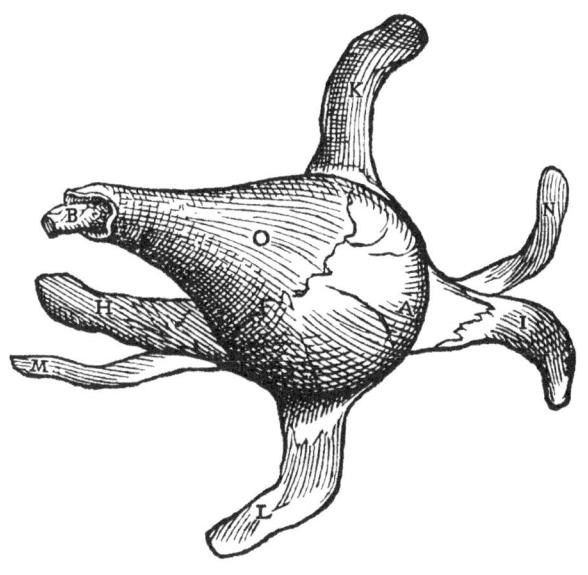

imagines a third muscle that would be responsible for elevation of the eyelid.[200]

My judgment about them is clear where I placed my account of the muscles of the eyelid.

When after six previous muscles Galen described the muscle of the eye which he sometimes perceived as several muscles and sometimes as a single one [*m. retractor bulbi*], he does not seem to me to show very well the function of the one by which the eye is kept in its place. In animals as also in humans, that muscle is inserted in the back of the eye more or less at its base, at the entry of the optic nerve. It ends in a circle as if at a point near its beginning, drawn over the optic nerve and is thus covered and coated by the six previous eye muscles and the fat lying beneath them, so that it nowhere touches the bone of the skull. I hear that some people are putting it about that I am describing bovine eyes rather than human[201] because they have seen that I always

---

[200] A passage in *De locis affectis*, 8.220.14ff., notes that both the eyelid muscles and their innervation are hard to see in large animals, but that there is one muscle that opens the eye and two that close it (8.221.4–7).

[201] In *De re anatomica* (1559, p. 216), Realdo Colombo wrote that like Galen before him Vesalius dissected an animal instead of a human eye.

bring bovine eyes to dissections and because the eyes illustrated in my book appear much larger than the eyes of a human. I have certainly always displayed bovine eyes in classes because human eyes outside the order of dissection[202] were too flabby and small. I would, however, also dissect human eyes whenever they were available. I did so more than once in private dissections when I judged it was permitted to look at [human] eyes soon afterward, but I saw no difference in the muscles. However, the pupil is much smaller in the human, nor are the green, blue, and very black colors to be seen on the inside of the uveal tunic [*tunica vasculosa bulbi, choroidea*]. Therefore I am still compelled to admit that Galen needs to be corrected in describing the use of the muscle that he enumerates so casually. ⁎

158   On account of the distinction of colors in the uveal tunic, I say that bovine eyes, like the brain of cattle and not humans, were what he dissected: the green, blue, and very black colors occurring in the inner region of this tunic but not the one that stands on the outside of the uveal tunic where this faces the cornea and the transparent tunic of the eye. Because of this color, we say humans have white, black, or grey eyes. That is the color in the uvea itself, whether the eye is seen still full of humors, or after they have been drained.

I would like careful students to pay special attention at the point where Galen writes about the temperaments of the eyes resulting from this distinction of colors. It is unlikely that those who consider Galen's fanciful thinking will fail to wonder at this Paradox of mine[203]

---

[202] On which see n. 179 above.

[203] The 1555 *Fabrica* (Bk. VII, ch. 14) explains the paradox as follows: "This color is never the result of an abundance of humors in the eye, a lack or thinness of humors, the accumulation or enlargement of the pupil, or the depth or shallowness of the eye, nor should it be thought the result of some similar cause. Though it may seem a great paradox to those who have filled entire books about eye color, I can approve no view except that the color of each person's eyes (however varied it may be, and inconsistent in either eye and sometimes even in the same eye) is always the same in the uvea, whether you examine it when the eye is intact, in what runs out of a severed eye, in the humors as a whole, or when the uveal tunic has been detached and removed from those adjacent by dissection. I therefore believe one may argue

(if they are held by love of the truth), unless we are willing to invent something about color, which is clearer or less clear in the pupil itself depending on its size. But it is easier to talk nonsense about this than to make true statements, so we may set aside the authors.

Returning to the muscles, from which my theme had digressed, I shall say nothing about the temporal muscle and the masseter; in describing these and telling their use, Galen was quite false in *De usu partium*, though in *De anatomicis administrationibus* he corrected himself and gave a fine description of these muscles without mentioning *De usu partium* – as if he hoped he would review that work some time. As it has happened in certain other places, the muscles that flex the first finger joints, the muscle [*m. popliteus*] hidden behind the knee, the muscle in the axilla of apes and dogs that is made of a fleshy membrane, and some others that are mentioned out of the usual order in *De usu partium* ✳ and sometimes interrupt the argument he is making as if he had only jotted them down in the margin, show and prove how casually the description of the muscle located behind the knee comes in. However, based upon what Galen himself says, it is clear elsewhere that these muscles were unknown to him when he wrote *De usu partium* and the earlier books of *De anatomicis administrationibus* which had perished in a fire.[204]

159

But these matters have to do with another place. I began this account so that you would not think me so negligent that I would not have observed statements made by Galen that are bad anatomy and scattered in his book about the use of the parts, concerning the muscles that move the lower jaw. Nevertheless, because I have come upon the mention of contrary and conflicting passages in Galen concerning

---

about eye color in the same way that one may argue about a black, swarthy, or white skin given to people in utero, or about red, blond, or black hair, as we see about the same kind of skin and eye color is observed in various nations."

[204] "In 192, a great fire burned down the Temple of Peace and many other buildings in the neighbourhood. The temple was a meeting place for intellectuals, and also served as a book repository and store. Galen lost all his copies of his own books in it, some of them irretrievably." Hankinson 2008, 21.

a description of muscular function, I cannot omit the muscle [*m. omohyoideus*] that I describe as implanted from the scapula into the hyoid bone and declare to be a private muscle belonging to that bone, as Galen also in *De usu partium*, when listing the muscles of that bone, particularly describes. However, in the same book, just as he does in *De anatomicis administrationibus*, he introduces this muscle as a mover of the scapula; this muscle therefore shows no trivial inconsistency in Galen: especially as it is a slender muscle, it cannot be counted in the number of those moving the scapula.[205]

What is the point of discussing muscles moving the head, that are conspicuous in so great a number? I am altogether certain that Galen was deluded in describing motions of the head. The muscle [*m. teres major*] originating from the lower rib of the scapula and ending in a long insertion below the head of the humerus where the first muscle of those moving the arm [*m. pectoralis major*] * is implanted and takes for itself the front of the breast, draws the arm backward in a motion which I identify as the first. It is by no means (although Galen said so) the author of rotation of the arm to the outside. From this it is immediately clear to what extent Galen perceived Nature's artifice in the muscle [*m. infraspinatus*] occupying the convex side of the scapula and attached by a large insertion to the head of the humerus and the ligaments of the joint here, and causing rotation.

The muscle [*m. latissimus dorsi*] drawing the arm downward, originating out of the back from the spine of the sixth thoracic verte-bra all the way to the lower part of the sacrum, is not inserted in the scapula; Galen wrongly gave it the function of moving the scapula. The muscle in simians and dogs that is made of a fleshy membrane and is not present in humans readily shows by the actual course of its fibers, which Galen describes accurately, that it does not perform the function Galen ascribes to it in a simple and direct downward motion.

---

[205] See p. 58 above for Vesalius' first mention of this topic. In *De usu partium* Galen described the omohyoid muscle as moving the shoulders forward or toward the neck: 3.592.18–593.1 (May, 1968, 374) and 4.140.5 (May, 1968, 618).

Likewise, Galen badly described the origin of the muscle [*m. pectoralis minor*] which I number the first of those moving the scapula;[206] it lies beneath the muscle drawing the arm to the chest and is inserted in the inner process of the scapula, not in the humerus nor in the ligament of the joint, particularly in humans. He also made a major error in describing its function when it was not enough for him to have written that the arm is adducted to the chest with the aid of this muscle but he went on to say that it performs its highest adduction toward the clavicle. In the first place, it does not move the arm; and since it makes its beginning much lower, from the second rib or a little lower than the first of the muscles moving the arm, which it brought from the clavicle and the sternum, it is quite ridiculous to think that this muscle of the scapula, even if it were inserted in the humerus, can adduct the arm higher ⁎ to the chest than the higher part of the first muscle moving the arm [*m. pectoralis major*]. I will not say how much lower it is inserted in the humerus and descends lower from its higher parts than the insertion would be of a muscle moving the scapula, wrongly ascribed to it by Galen.

In the muscles moving the thorax, I shall bypass the diaphragm itself along with other muscles and consider only the intercostal muscles because their use, unknown by Galen, makes me aware that he wrote many things that he had never seen or dissected but had conceived only in his imagination out of false principles. Everyone who has not surrendered to the authorities but believes in the truth will agree with me and consider it evident that Galen did not know the function of the intercostals. As he himself is more than once inconsistent and often varies both in his descriptions of things and in assigning uses and functions, so too we find his opinion about the function of the intercostal muscles is unsettled. He everywhere teaches that these muscles relax and compress or tighten the thorax, or are in charge of

---

[206] This argument was made in *Fabrica* II ch. 23 in the section headed "The muscle that moves the scapula forward is mistaken by Galen for a mover of the arm" (pp. 263f.).

inspiration and expiration. But in one place he writes that the inner intercostals open the thorax and separate the ribs while the outer intercostals compress and tighten it, while in another, by contrast, he declares that the thorax is opened and the ribs separated from each other by the outer intercostals and the ribs are gathered together and the thorax tightened by the inner. Since this conflicting opinion presents itself, it is easy to believe it is one and the same, taking refuge in errors of the texts, since it is easy to write "exterior" for "interior." But however we now imagine Galen's opinion to be consistent, it must always be admitted that Galen made either the inner or the outer intercostal muscles ✳ the authors of opening of the thorax and of separation of the ribs from each other; but I should have thought both the outer and the inner muscles control the drawing together of the ribs and the tightening and narrowing of the thorax.

I have never been able to invent an explanation of how soft bodies in the spaces between hard ones should be able to be drawn together (which is the chief property of muscles) in such a way as to force the hard bodies to draw apart from each other or separate, and not come together. Nor has anyone so far tried to contrive me a way – in the presence or absence of a cadaver – in which Galen could have proven his opinion, especially after he had closely inspected the nature of intercostal muscles. So many reasons present themselves why they all act to tighten the thorax. I have considered them in dead and living bodies with all possible diligence, earnestly asking those in attendance that if they disagreed with my opinion to give a reason for their opinion (other than the authority of Galen) why it should [not] yield to mine.

It is an embarrassment to the name of Galen, whom we consider our common preceptor and our guide everywhere through the entire art of medicine, that he erred in assigning the task of these muscles, and labors under such suspicion among us because hereafter we will not dare trust him without the best reasons. He tells of many vivisections that hinge on the function of the intercostals, so it is all too clearly known that he conceived of them with little intelligence if he did not know their function. But there are several such instances which seem unskillful in other ways and testify that they were more imagined than tested by experience.

What shall I say about the use of muscles moving the back as mentioned by Galen? He paid no attention to four major muscles moving the back and passed them over ✳ as if they did not exist. I think nobody is so stupid that because of ignorance of those muscles he does not believe the function of other muscles as well is wrongly described. Yet in numbering them and describing their function Galen was so succinct that very few errors can be counted therefrom in his account of their uses. However, in *De usu partium* he wrote falsely that the muscle flexing the third joints of the four fingers is also inserted by its tendons into the first bones; he added that those bones as well are flexed by this muscle. But when he wrote *De anatomicis administrationibus* and considered on the basis of a more careful dissection that this insertion is not in the nature of things,[207] he attributed the same function to the muscle just named. But he attributed flexion to the transverse ligaments by which the tendons are contained above the first bones. Since, therefore, Galen corrected himself, we must accept the opinion here that is closer to the truth, as we always do elsewhere.

So Galen stated that the first digital bones are flexed with the aid of the transverse ligaments, which are authors of flexion of the third bone. That seems to me at odds with the truth, on the basis of the strength of the first bone's flexion. For the same reason why the tendons flexing the third joints flex the first, those that flex the second joints would also be the ones flexing the first because they are contained in the same ligaments in the first joints as those that flex the third joints. Moreover, the first joints each have two muscles that belong to them, by which they perform flexion; but the first joints

163

---

[207] Vesalius described Galen's change of mind in *Fabrica* II ch. 60 (p. 350, misnumbered 250).

are more weakly flexed than the third. Therefore even under the fixed authority of Galen it was neglected by him that the first joints are
164 flexed by the same tendons as are inserted in the second joints, ✳ and the second joints also get their flexion from those tendons that are implanted in the third, since they are surrounded by the transverse ligament no less in the second joint than in the third.

Galen also neglected to say that the wrist should be flexed by these tendons, since they are enclosed by the transverse ligament in its groove. So too there are many bones in the foot regarding which Galen was lacking if he also wanted bones in which tendons are not implanted to be moved by transverse ligaments. Granted, however, that when describing the function of the tendons moving the toes he was influenced by this reasoning and stated that the tendons controlling flexion of the third joints also move the first and second; but at that point everything in Galen is so obscure, and what is written in *De anatomicis administrationibus* agrees so little with what we read in *De usu partium*, that in this part I have never undertaken to decide anything definite about his opinion.

When Galen also ascribes to the muscles moving the wrist the function of pronating and supinating the wrist, he is no less mistaken than when he states that the wrist is pronated and supinated by its own motion. It cannot perform that motion; likewise, the muscles moving the wrist also cannot be in charge of that movement. Because in his account of the muscles moving the forearm in *De anatomicis administrationibus* Galen is so inconsistent with the one contained in *De usu partium*, and because everything in the former book is more true than what is in the latter, I shall leave out what is quite useless in *De usu partium* about the crossing of fibers and many other topics lest someone think I take pleasure in chiding Galen and believe I do not, like everyone in the medical schools, grieve deeply about so great an author.
165 But I cannot omit to say what ✳ he got wrong when he described those muscles in *De anatomicis administrationibus*, correctly, at least, about apes, in describing their function: that is when he ascribed

to those muscles, which run altogether straight and make no crossing or intersection of fibers, such oblique motions.[208] It is not doubted that the forearm is not moved by its own motion beyond straight flexion and extension. For if someone flexes the forearm now to the chest and now to the tip of the shoulder, the diversity of flexion depends upon rotation of the humerus and not upon the elbow joint, which is attached by ginglymus and is capable only of simple motion. Thus the shoulder joint confused Galen more than slightly in his account of the muscles moving the forearm.

How uselessly he occupied himself in explaining the use of the ligament inserted from the ulna into the wrist is clearer from the actual makeup of the joint of the wrist to the forearm and the independent motion of the wrist, than requires that I should linger refuting Galen's inaccurate description of that ligament's function.

In the muscle controlling the neck of the bladder, because of symptoms in the bladder and its neck that have recently come to my attention in sick persons, I would like to understand more than the smallest amount so that however often I have dissected the neck of the bladder, I could subject it to a thorough examination at each first onset of symptoms. But so far as concerns the present business, I am astonished at how Galen teaches that the same muscle assists the promptness of retaining and expelling urine as its particular function; but I believe that the better urine is removed when pressed by the circular fibers of the bladder and especially by the transverse septum with the abdominal muscles, the more the muscle of the neck of the bladder opens and the more it relaxes its own fibers. Nothing else occurred to me which I could think is the cause of voluntary retention of urine ✳ in addition to the gathering onto itself of this muscle. What Galen says about the narrowness of the neck of the bladder and the urinary passage, and its length and turnings, does not have any

166

---

[208] For these remarks about Galen's errors in describing the muscles flexing and extending the forearm, see *Fabrica* II ch. 46 (pp. 316–19, misnumbered 216–19).

relevance in women, whose urethra is quite short and runs straight downward to its insertion into the neck of the uterus [vagina].[209]

Now as the muscles moving the tibia provided a rich abundance of discrepancies by which it is shown that Galen dissected apes and did not see humans, and as the same muscles have put forward such a vast heap of untrue and conflicting descriptions that I have not tried to recite them, so too the muscles moving the tibia present themselves to those who look uncommonly hard for Galen's diligence in his account of their actions and functions. First in *De usu partium*, where the muscles moving the tibia were described outside the number of the Ancients and almost solely from his imagination, Galen taught that the lower leg is extended and then flexed but in a triple variety: two oblique flexions of which one is inward and the other outward, and one straight, or halfway between the oblique movements.[210] In addition, he attributed sideways motions to the tibia, by which it is moved inward and outward. There was also an oblique motion, which he assigns superfluously to the tibia in *De anatomicis administrationibus*, by which we move the tibia sideways when raising it as if we were moving it toward the other tibia. Galen ascribed muscles to these motions, quite contradicting himself in *De anatomicis administrationibus* and *De usu partium*. But I am not going to cite the difference of his opinions, since I wish to show that the tibia moves quite differently than Galen says. This part of the leg moves only by simple extension, which goes as far as an acute angle, and has no oblique movement, ✳ as its joint with the femur made by ginglymus shows quite fitly. The tibia does not perform the lateral and oblique movements declared by Galen in its primary motion but only in secondary movement, undergoing every kind of movement at the

---

[209] See n. 189 above. Vesalius does not distinguish between the vagina and its vestibule, where the urethra ends, just as he does not distinguish between the cervix of the uterus and the vagina. The vagina was not distinguished as a separate entity until Falloppio's *Observationes anatomicae* (1561).

[210] "Just as [Galen] made up false motions of the tibia, in the same way he handed down to us a muscle fashioned by his imagination." *Fabrica* II ch. 53 (p. 335 misnumbered 235).

femur in the hip bone, where there is enarthrosis, like the movement of the humerus at the scapula. That this is the means by which the tibia is moved to the inside or outside when we flex it is so certain that nobody so informed could be in doubt if he has conscientiously taken into consideration the joints and motions of the bones.

Since therefore my view about the motions of the tibia is so different from Galen's, it is also clear how I disagree with him in assigning the function of muscles moving the tibia. But I am most of all surprised that the motions of the tibia were not precisely or truly observed by Galen, since the function of the muscle [*m. biceps femoris*] which he numbers fourth of those moving the tibia in *De anatomicis administrationibus* is described otherwise in *De usu partium*, where he places it seventh of those moving the tibia, stating that the tibia is abducted to the outside together with flexion, and is rolled somewhat inward; but in *De anatomicis administrationibus* it is attested that the entire tibia is moved by it in a simple motion to the outside. I am persuaded that because Galen mentions this muscle often, and always calls it a wide muscle, believing that it had been torn away in a runner in a competition,[211] he diligently undertook an examination of that muscle's function in apes, and in this way easily observed that it moved the tibia only in a simple motion. But since this one reserves a more oblique course than the third muscle [*m. semitendinosus*], which Galen says originates from the epiphysis of the hip bone, runs along the inner head of the femur, and is inserted in the front of the tibia, he wrongly determined that it is the most oblique of all the muscles moving the tibia. A more oblique course is without doubt that of the fourth muscle [*m. biceps femoris*], as it runs from the same epiphysis

---

[211] Galen's account of the runner is in *De anat. adm.* 2.298.17 ff.: "In the case of a certain excellent runner, we saw this muscle [*biceps femoris*] ruptured about the middle while the man was racing. After that its place was empty and hollow, for the parts of the torn muscle had moved, the upper being pulled toward the origin, the lower toward the tibia. When pain and inflammation had subsided, walking did him no harm and, taking heart, he began running again. Feeling none the worse for this, he actually restarted racing and was again victorious." (tr. Singer 1956, 39 f.).

Detail of the 11th Table of Muscles in Fabrica II. The two movers of the lower leg most disputed with Galen are the semimembranosus (Ψ), whose origin is marked ν; the tendon near its insertion in the front of the tibia is marked ο. The semitendinosus (ς) has been cut away from its origin (λ) and hangs from its insertion in the tibia. The other hanging muscle (σ) is the long head of the biceps femoris.

outward to the outside part of the femur and so to its outer head and is also inserted in the tibia, having the same course in simians and humans, though in its method of composition and appearance it is significantly different in the former and the latter.

When Galen argues at such length in *De anatomicis administrationibus* about the fifth muscle moving the tibia [*m. semimembranosus*] and is so inconsistent in the same page, I wish he had not dissected some other muscle along with the third [*m. semitendinosus*] and therefore missed the muscle that I call the fifth moving the tibia. In such a case there would have been no disagreement between myself and Galen about the function of the fifth muscle.[212]

---

[212] This disagreement is explained in *Fabrica* II ch. 53 (p. 332, misnumbered 232) in the section titled "What muscle Galen calls the fifth, and which one is truly the fifth."

But these warnings must be given only in passing, lest I be forced to descend into a lengthy description of the muscles and recount what I clearly stated on my part in *De humani corporis fabrica*. I wondered there in his description of the sixth muscle moving the tibia [*m. tensor fasciae latae*] that Galen did not consider how by the unique workmanship of Nature it acted like a membrane to contain the muscles occupying the femur, and surrounded them like a transverse ligament to keep them from being changed in any way; but Galen in *De usu partium* imagined something in place of this sixth muscle when he contrived to say that a muscle abducting the tibia to the outside originates from the outermost side of the hip bone. Perhaps somebody will argue that Galen meant the ilium here, unlike the rest of his account of the muscles. But this will again be justly seen to pertain rather to descriptions of the muscles than to an explanation of their use (as if I should likewise argue that Galen had wrongly attested, also in *De anatomicis administrationibus*, that the ninth muscle moving the tibia [*m. rectus femoris*] originates from the femur).

I shall therefore say nothing more about muscles here if I have first given notice that I have no reason, based upon close examination of the muscle [*m. popliteus*] hidden behind the knee, to believe that any function of bending the back of the knee or the knee can be ascribed to it – though Galen sometimes appears to believe that this is the sole author of flexion of that joint, or at least that it has the special power of setting it in motion; he regarded it with excessive favor perhaps because he was the first to observe it or because the Ancients had not counted it among the muscles that create motions of the tibia.

Because something was briefly set down about the muscles moving the digits of the feet when I spoke about the muscles of the fingers, I will pass them over for the present and add a few things about other organs.

169

The 1555 *Fabrica* contains an extensively rewritten description of the *m. semimembranosus* and Galen's errors concerning it.

Places collected from the description of veins, arteries, and nerves where it is known that Galen consistently assigned incorrect uses and actions

It is not entirely to my liking that Galen attributed to the portal vein the power and faculty of preparing juice to be supplied from the intestines to the liver in a way that is extremely similar to the function of the liver itself. For though I am not forgetful of our common opinion about the settling of urine, I cannot persuade myself how the power of making blood can exist in a white, thin, membranous body; and if some power of digestion were present in a vein besides that which all the parts possess for preparing their own nourishment, it would certainly have created a white humor rather than blood. If someone called my attention to the proximity of the liver to the whole length of the vena cava, I would like him to see that Galen was so forgetful in *De placitis Hippocratis et Platonis* of what he had written elsewhere and so excited about arguing with Aristotle that he denied the power of sanguification in the very short interval of the vena cava that is seen between the liver and the heart.[213]

170   That paradox has a bearing on branches of the portal vein extending to the place where I stated that thick, melancholic ✳ blood flows out of the anus at intervals, more through those branches than through offshoots of the vena cava. I made that statement in the Epistle that states, among other things, chiefly that it makes little or no difference which vein of the forearm is opened in the lateral disease; or if

---

[213] See *Fabrica* III ch. 5 (p. 367, misnumbered 267): "I can assign no power of sanguification to the membranous and leathery body of a vein; if there were, it would surely make a white blood, just as substances transformed by the stomach resemble a milky color like the white substance of the stomach. Clearly we do not see, as in the liver, that it is red like congealed blood. I would refute this oft-repeated view of Galen at greater length, were it not that Galen himself changed his mind in the sixth book of *De placitis Hippocratis et Platonis*, forgetting himself so as more sharply to dispute Aristotle." Galen said in *De usu partium* that the veins have a role in sanguification.

it makes any difference, that the vein of the right forearm should be cut, whatever side or whichever part of the thorax the inflammation attacks.[214]

In the description of nerves, as on many other topics, I am quite surprised at the question commonly put in the medical schools which is taken from the first book of *De locorum affectorum notitia*[215] where it is disputed at length and with all kinds of arguments how movement can be lost while sensation remains, and how when motion is preserved sensation is lost, the entire disputation being made always about the ring finger and little finger. Here they all say first that a greater amount of animal spirit is required for motion, and therefore when that spirit is lacking it can easily happen that when motion is lost sensation is still to some degree unaffected. But the reason does not so readily appear how motion survives after sense has been lost. Galen appears to have made a distinction so that he could say that some nerve twigs are responsible for motion and some for sensation, as if a nerve occurred that lacked a sense of touch.[216] I can for no reason subscribe to this opinion, especially because for one trained in the dissection of nerves and muscles there is no need to take refuge in that false axiom to teach that motion is preserved without sensation, especially in the fingers about which Galen made his argument. If one knows that what I count the fourth nerve [*n. radialis*] of those that enter the arm provides twigs not only to the muscles that most extend the fingers but also to many other muscles, he will also be aware that it does not go to the ring finger and little

---

[214]   See n. 184 above for the afflictions identified as *dolor lateralis*. The 1539 Venesection Letter in fact states clearly that "In all inflammations of the sides of the thorax or of the thoracic vertebrae, the right axillary [vein] must be cut" (*Venesection Letter*, p. 56, tr. Saunders & O'Malley 1947, 82) – though it appears here that Vesalius has changed his mind.

[215]   Also known as *De locis affectis*.

[216]   It was in fact Herophilus of Chalcedon (c. 330–260 BC) who first distinguished motor from sensory nerves.

finger ✳ as it does to the others with offshoots to their outer side, but that those two fingers receive twigs on both the inside and the outside from the fifth nerve [*n. ulnaris*] entering the arm. He would be able to damage the fourth nerve in the back of the upper arm or in the elegant complex of nerves near the sides of the

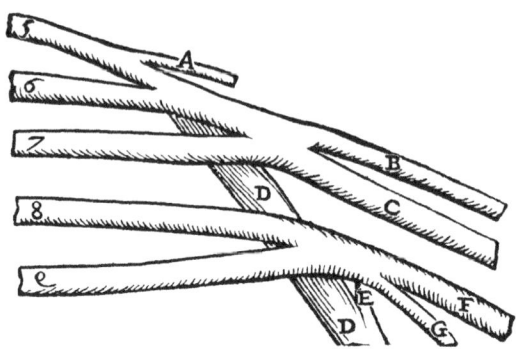

The "elegant complex of nerves" leading to the arm from the cervical vertebrae, from the beginning of Fabrica IV ch. 14. The radial nerve is labeled D; F is the ulnar nerve.

cervical vertebrae, or learn that it had been bruised or crushed while the fifth is still undamaged: he would immediately notice that sensation had been lost to the little finger and the one next to it, but their extension is not spoiled. Again, if he imagined the fourth nerve unhurt but the fifth damaged, the function of the muscles extending the fingers will not be lost but sense in the little finger and the ring finger will be absent. The truth of this proposition (if it ever turns out to be true) must be sought from careful dissection rather than an imagination whereby we suppose that nerves lack a sense of touch contrary to the nature of nerves.

But whenever this widespread and too common question is raised we must think about the flow of humors in the nerves rather than its division, since the latter would deprive the entire nerve of sense while the flow of humors sometimes indicates the first damage in the farthest twigs of a nerve, however deep the nerve itself stands.

But it is not my intention to write you, my dear Joachim, about such distinctions that must be declared in academia for us to dispute for the sake of sophisms: you have enough erudition and sharp judgement to know immediately the opinion to which I am speaking and to understand that my effort is not for a beginner in our craft.

A description of some things that are contained in the peritoneum

To say nothing about the function which Galen provides to the peritoneum of preventing the intestines from coming between the muscles that lie upon them, I cannot imagine how the peritoneum also ✳ performs the functions of the nearby transverse septum as an aid in expelling feces more than as a skin cloaking the abdomen or the omentum, or as fat lying beneath the skin. While Galen ascribed this function to the peritoneum, he should also have considered how by the same token he should have attributed to the membrane covering the ribs the capacity of compressing and narrowing the thorax.

172

Galen and everyone who wrote about anatomy after him attribute a glandular body to the lower orifice of the stomach by which they think it is contracted;[217] some of them also add to what they wrote that this body surrounds the orifice like a kind of muscle, and has a function similar to the neck of the bladder or the muscle of the anus. I am certainly aware of a long glandule next to the duodenum, similar in substance to the body that is placed on the lower membrane of the omentum and takes its name from a type of meat and its good

---

[217] The pancreas, whose alleged function was previously discussed on p. 133. In the 7th ch. of Bk. 4 in *De usu partium* Galen writes "In many animals something of a glandular nature is found at this point [the pylorus], which increases the constriction, particularly when the stomach exercises its retentive faculty and is actively engaged in digestion, gathering itself together, contracting, and clasping and compressing its contents (3.280.18–281.5, tr. May 1968, 211, who notes "This is a puzzling statement and there is nothing to explain it in Galen's other works. Could he have been thinking of the head of the pancreas? But if so, why the qualification 'in many animals'? In the rhesus monkey the folds of the mucosa in the pyloric canal are high and distorted ... and this feature may be what was in his mind," etc.).

looks.[218] But it is perfectly clear to me that this glandular body is not responsible for closing or opening the orifice of the stomach, but is an elegant support and bolster for the vessels running lengthwise along the intestine; it would therefore not seem worthy of an anatomist to put this gland in charge of excretion which is natural and not subject to our will.

I do not know whether it was taken from Galen somewhere that the upper orifice of the stomach is closed by the transverse septum by means of its foramen by which it transmits the esophagus, because people declare that it also prevents the upward regress of food; they do not observe that if their statement were true, that foramen of the septum would no less block the entry of food into the stomach. It is easily known that we have no motion in the orifice of the stomach that depends upon our will.

Concerning the function of the spleen, the natural purging of impure and black blood into the stomach ✳ and of what would become single products of digestion (as we believe happens with yellow bile), and about constriction of the upper mouth of the stomach, I am no less uncertain than I was when I published my book *De humani corporis fabrica*. I am perfectly willing to have it blamed on my ignorance that I do not find the passage or vein that spews the black juice[219] into the upper orifice of the stomach rather than into the rest of the left side of the stomach. I do not say this as if I thought a long discourse should be commenced here about the functions of the spleen, but so you would understand that I am still in the same way of thinking and still more that I have doubts about the common

---

[218] Gk. πάγκρεας, "all meat" and καλλίκρεας "beautiful meat" because it was favored as a food; cf. Engl. "sweetbread." On this organ, see *Fabrica* V ch. 3, the section titled "Glandular flesh standing not far from the lower orifice" (p. 491, misnumbered 391).

[219] The mythical black bile that accounts for melancholic personalities and constitutes one of the four humors. Vesalius has reverted to this subject several times already in this work. See pp. 133, 135, 138, and 221 and notes 174, and 181. On the spleen, see *Fabrica* V ch. 9.

opinion of physicians even since I inspected the spleen of Belloarmato of Siena, the great jurisconsult at Pisa, which beyond doubt had for a long time performed the function of his liver.[220] He had greeted me in a bookshop to which I had gone in the afternoon with my students when my lecture on dissection was completed, and he had been asked some questions about his health, which had been poor for many years. In that conversation I mentioned obstruction in passages of the liver, the gall bladder, and finally the spleen as well; he said that the next day he would come to the anatomy and closely observe the organs whose construction was due to be explained. When he went home from the bookshop and occupied himself with his studies for a few hours, soon after beginning his supper he was overcome with a strange weakness of body and shortness of breath. After several remedies which were considered appropriate for the flow of bile into the stomach were tried by doctors who knew him, he expired. Because his body was to be transported to Siena and placed in the family tomb, his family and friends asked a surgeon that his organs be removed. ✳

When he proposed this task to me in the beginning of the morning, I was quite eager to learn the cause of such a sudden and unexpected death in such a famous man. When I dissected him, I immediately found all the blood in his body to be still quite warm and to have collected as I have sometimes seen water collected there, in the peritoneal cavity, an accumulation under the skin which we know takes its name from a skin water bottle.[221] A hardened abscess in the stem of the portal vein had provided the occasion for this flow of blood; the abscess had suppurated in one place and broken, giving the blood a path. So as soon as I had removed the brain and all the

174

---

[220] On Marcantonio Belloarmato and the ailments that led to his death, see O'Malley 1964, 202, 451f.

[221] Lat. *uter*. Cf. *Fabrica* V ch. 2: "Like wine sacks [*utres*] made from skins that are usually hairy and rough on the outside and smooth and even on the inside, the peritoneum appears rough and fibrous on the outside to adhere better everywhere to muscles, in exactly the way we say that membranes attached to each other are uneven." The

viscera, and thus prepared the remaining body so I believed it would be less tainted and freer from putrefaction, I had his liver, gall bladder, stomach, and spleen transported with me to the university to exhibit as a great proof of bad health.

We saw that the liver was quite pale, like the inflated lung of a pig or a dog. Its surface was not smooth but quite uneven and roughened by many smooth tubercles. On the hollow side of the liver, all the branches of the portal vein were plainly visible after the substance of the liver had been fully retracted or pulled away from the lower area of branches. The anterior part of the liver and the entire left side were hardened like a stone, but the back where the stem of the vena cava stands was quite soft and appeared damaged by decay.

The gall bladder was also paler than normal and contained eighteen small stones, quite smooth and shaped like a triangle with sides and surfaces that were everywhere equal, green and blackish. When dried they appeared more ashen, in size like chickpeas.

175 The spleen, for whose sake I am chiefly writing this, ✳ was quite large, displaced to the front of the body, soft, and a little whiter than a natural spleen. To put it simply, no one was found in so large a crowd of extremely learned spectators who would not immediately state that the spleen had indubitably functioned in this man in the role of the liver, and his liver appeared for a long time to have had no share in the role of sanguification. Meanwhile, the spleen had grown to the size of the liver and was modified also sufficiently to produce blood, so far as we could gather, in the degraded constitution of his body. And since its veins extending to the stomach from the vessels closest to the spleen were quite large, I also advised the students to observe carefully whether a vein was inserted as far as the mouth of the stomach. But in the vein that came closer than all the others to that orifice, we found no difference from that of other humans except in size. Like all the others connected to the left side of the stomach, it was by far the largest.

---

term survives in modern anatomy only as the diminutive "utricle," a minute pouch in the prostate.

I also dissected Prospero Martello, the Florentine patrician (who for many years had suffered from jaundice and like Belloarmato had died a sudden and unexpected death). When I was about to leave Florence and was riding past his palace in company with Francesco Campana, chief private secretary of the Duke, I was called upon by certain gentlemen to investigate the cause of death with a number of surgeons who had already begun the task. Death had been caused principally by an injection of bile into the stomach, which was swollen with pure bile, accompanied by hard swelling of the liver and its contraction or condensation into one. The spleen, however, was softer and larger than normal, and appeared to have functioned in preparing blood; but the gall bladder was easily the size of two fists and full of something like tiny pebbles, ✳ which were connected together and were very like grains or seeds of millet, or rather the rough surface of the common tutty[222] of the pharmacies. Wherever I opened the veins, I found nothing but very thick bile, and indeed the fluid in the arteries stained my hands no less than the bile itself. I have reviewed these facts so you may understand why I am in doubt about the function of the spleen and why nothing is ready to be revised in my account of it in my book. As the Gods love me, I therefore have found nothing written there, especially where I have cited Galen, where I believe anything unreasonable has been included, however carefully I reviewed them after publication and the Geldric war[223] from which I returned to Italy and performed public anatomies at Padua, Pisa, and to a degree at Bologna when I was about to set out from there to Pisa. I could not fail at the urging of friends to dissect several parts which were ready at hand there in an anatomy when the business had run late into the night; Buccaferreo of loving memory and Albius (to

176

---

[222] Medieval Lat. *tutia*, an oxide of zinc used medicinally in astringent ointments and lotions (OED).

[223] Gelderland, a duchy of the Holy Roman Empire, became part of the Habsburg Netherlands in 7 September 1543 following the siege of Düren by Charles V and the Treaty of Venloo.

both of whom I owe much for their singular candor toward me) were eager for me to do this, together with a large number of students who were present.

I had also examined the organs of the Prince of Orange, Seigneur de Hallewyn,[224] and a number of others whose bodies were injured in bombardments and had to be removed by someone from our army. I never thought the opportunity would be lacking to examine the things of which I shall be reminded, though I should rather persuade the disciples of Sylvius and Sylvius himself to consider carefully everything I say before attempting to disprove or condemn them from my book alone – lest they prematurely remove the opportunity for students to investigate the truth. I am not one to believe anything human is alien to me,[225] nor am I unaware ✳ of the steps in which I have learned the fabric of the human body by teaching others and by writing. In this regard I acknowledge no preceptor whatever, nor do I believe if I say so that Sylvius will later take offense, if while he attests that Galen made no error I loudly contradict him so many times. Sylvius also writes that no matter how much attention he pays me he will not keep me as a friend unless I hold the same opinion he does about our common preceptor Galen. No one should therefore expect great praise from me for Sylvius nor wish me ill because I do not consider him magnificent among the professors of anatomy. I must acknowledge that all who follow Galen today have the same view as Sylvius and have passed it along in their books, if they have not also sometimes misrepresented Galen, never noticing an inconsistent place in his writings. As least on the basis of what I have written here, it is quite clear that many discrepancies exist in Galen's books.

---

[224] Jean III van Halewyn (1510–1544), casualty in the siege of Saint-Dizier, France, 23 July 1544. See O'Malley 1964, 206, 452 n. 80. Vesalius misspells his title *Arangia* and his name *Haluin*.

[225] From Terence's famous quotation (*Heauton Timorumenos* 77), itself a tag from Menander, *Homo sum: humani nil a me alienum puto*.

I could not therefore be sufficiently astonished by the letter of Dryander shown to me at Cologne by our common friend the prominent doctor Ioannes Eck when I recently traveled there,[226] in which he complained that I had made no list of the celebrated professors of anatomy of our time, and that in addition to a number of others (I do not know whether they were ever born) I had not mentioned Ioannes Guinter in this way as my preceptor:[227] I certainly respect him among many names, and for his public writings as a preceptor of medicine. But I wish as many cuts made in me as I have seen him make in a human or another animal (except at the dining table). I do not think Guinter objects to that since it is known to him and many others whether he owes anything to me in this part of our craft, if in fact he claims anything for himself in the method of dissection beyond the common books of Galen. ∗

But Dryander brings forward this person, perhaps thinking he himself has been wronged because he was not named in a list of professors and because he is unhappy that I censured him when I wrote that someone at Marburg was publishing figures taken from the books of others wherever they were, as if they were his own.[228] I never thought that Dryander, of whom I previously held a different opinion, was publishing the anatomical books which I believed printers had brought

178

---

[226] August 9–17, 1545; see O'Malley 1964, 210, 221, 453 n. 104. Eck was a physician from the Netherlands who practiced in Cologne and had a special interest in botany.

[227] Johan Guinther (or Winter) of Andernach (1505–1574) lectured on anatomy at Paris when Vesalius studied there and was assisted by Vesalius. His *Institutiones anatomicae* (1536) anticipated Vesalius' polemic against physicians who were ignorant of anatomy, but its anatomical content was purely Galenic. See O'Malley 1964, 54ff.

[228] Identified by Choulant 1920, 148 f. and Cushing 1962, 28–32 as Johannes Dryander, alluded to in Bk. 2 ch. 7 as a "certain mathematician" who misrepresented the saw best suited for cutting the bone of the skull, and more courteously by name in Bk. 5 ch. 4. He was appointed professor of mathematics and medicine at Marburg in 1535, and his *Anatomia capitis humani* (Marburg, 1536, 1537) was one of the first illustrated anatomy books. In 1541 he had plagiarized the illustrations in Vesalius *Tabulae sex* (1538). He is mentioned as a plagiarist without being named in Vesalius' Letter to Oporinus at the beginning of the 1543 *Fabrica* as one "who is still indiscriminately compiling pictures from other people's books everywhere and publishing books of

out in both Latin and German for the sake of sordid cash, borrowing his name (to make their own work more salable). And if Dryander had been the author of those books (as even he now persuades me he was) and gave me no reason why a list of celebrated anatomists of our time should have been put in the preface of my book when he himself should be content with common censure, I too must likewise endure it that like many others he resents Vesalius for the things many have now written about me and rejoices that Cornarius[229] will soon correct Galen and Aristotle in all passages where I have criticized them. Dryander remarked in the same letter to Eck what needed to be done especially by one who is eager to be numbered among anatomists and surgeons. Cornarius should abandon the labors by which he is carrying forward their common interests, especially since this correction of Galen is not to be sought from his books or those of others, but from the diligent and careful dissection of humans, simians, and certain other animals. Nor is it sufficient to occupy oneself in speaking ill of someone or ridiculing the efforts of others and to detract equally from one's own and others' glory (while so few are laboring to do credit to their studies), when one should rather be working up a sweat in common efforts at the truth, and believing that we too were born human. *

179   Something in the vast art of medicine may be present in us, as well as a faculty of discovery, if we are more strongly held by a desire for truth than for calumniating others.

With these thoughts, I regret the fate of Cornarius and Fuchs[230] that they have damaged their otherwise celebrated name when they

---

that kind at Marburg and Frankfurt." Vesalius referred obliquely to Dryander again near the end of the 1555 *Fabrica* I ch. 5.

229 "Primarily a medical philologist who edited a number of classical medical texts including works of Hippocrates, Galen, Aetius, and Paul of Aegina and belonged to that group of physicians who, because of their strong and constant belief in the validity of classical authority, were naturally unsympathetic to Vesalius's purpose." O'Malley 1964, 456 nn. 178 and 179.

230 Leonhart Fuchs (1501–1566), German physician and author of *De historia stirpium commentarii insignes* (Basel, 1542), which made him one of the founding fathers of

harass each other with such wrangling when they should judge each other a singular credit to Germany abroad, and proceed to exhaust among themselves hostile energies that should have been employed in praises of Germans and Belgians.

Since I blame this behavior on the unhappy failings of mortals and believe we ourselves open the window to such reproaches upon ourselves, lest I be distracted by another ignoble passion and let myself digress further, I shall return to Galen. When he inquires into the cause of the beginning of the vessels delivering material to the left testicle (both of which he stated inaccurately originate from the vessels that go transversely to the left kidney), and asserts that all the stimulation and pleasure in the ejaculation of semen depends upon the large amount of serous humor which these vessels supply, he seems to me not to have observed that men whose left testicle has for some reason been severed have the same experience in coitus as those whose right testicle we know has been removed. It is surprising that in urinating we do not feel the stimulation that we feel in the ejection of semen; however, where we experience the strength of the semen in the tip of the glans and the perineum, the same path is made for semen and urine; therefore when a man is healthy this passage is no more disturbed by the saltiness of urine than the bladder. And since we urinate sometimes when the penis is more flaccid and sometimes more rigid, we do not have recourse to the heating of the penis making it more sensitive to the quality of urine. Those beginning to suffer from an involuntary flow of semen perceive the excretion of semen with the itch of desire and stimulation even if ejaculation takes place from a flaccid and lowered penis.

From the description of parts located in the thorax and skull

The tip of the heart is quite thick and fleshy not to prevent it from being moved into the breast bone and damaged during contractions

---

botany. He and Cornarius carried on a well-known feud, as this remark of Vesalius shows.

of the heart, since it is clear that it is not bruised in this way; in fact, the nature and combination of fibers and flesh of the left ventricle show clearly enough why the tip of the heart turned out to be so thick and fleshy.

I wish Galen had observed in the ventricles of the heart that nearly the entire substance of the heart is taken into the makeup of the left ventricle, while the right is attached to the left like a crescent moon. The substance of the heart thus forms the right side of the left ventricle so that it protrudes to the right and keeps the same circumference and periphery as the remaining surface of the left ventricle.

But if I brought many of these things into consideration, an end to my writing would not appear quickly in my account of functions, and I must now hurry to conclude if I am to add that I am not entirely satisfied with Galen's reason why he determined that the cerebrum is in two parts: so that if one side of it was damaged the other would perform its function, in the same way that if one ear is stopped up we can still hear or if one eye is blinded we will still see with the other. Since the cerebrum is continuous for such a large area at its base, and then in its other bodies which are called by their own names, Galen should have mentioned the division of the cerebrum for suitable nutrition, as I have sufficiently shown that the convolutions of the cerebrum were made for its nutrition. It is clear how Galen thought about the uses of the cerebral ventricles when in discussing the excellence of the ventricles in *De placitis Hippocratis et Platonis* he says the exact opposite of what he said in *De usu partium.*[231] *

181   If one has observed the processes of the cerebellum [*vermes cerebelli*] which we compared to worms, and then makes an examination

---

[231]   Galen "taught in the third book of *De placitis Hippocratis et Platonis* that the middle ventricle was the main one, while in *De usu partium* it was the rear ventricle." (*Fabrica* 7 ch. 6).

of the nature of the cerebral ventricles, Galen will seem ridiculous for writing that the processes imitating a worm are responsible for opening and closing the cerebral ventricle. As the function of these processes is badly assigned, so also Galen is not to be believed about the tendons that he says contain the wormlike process on each side and prevent it from slipping to the sides when it is contracted or extended.

Many such remarks occur in *De usu partium* which indeed seem elegantly written and appear to set forth the great intelligence of the supreme Maker. But when one considers everything on the basis of careful dissection, he will not be able to wonder sufficiently how Galen like a Prometheus was able to invent a fabric of the body and uses of its parts which are quite foreign to the constitution of man. Likewise, whatever was said by Galen about the drainage of phlegm through bones and the organ of smell (as I explained earlier about his untrue descriptions) is known to conflict with the truth.

Some invalid anatomical proofs of Galen are mentioned

Next we shall weigh Galen's rigor in proofs and demonstrations, citing several proofs as examples (as we did before in other familiar places). Because there is little doubt that fat attaches to ligaments and membranes, and everything altered by the task of nutrition in our body assumes the color of what changes it, I believe that fat becomes white. This is not because it is like air; for otherwise membranes, tendons, ligaments, and still more, bones, would be black, being the driest, most terrestrial, and least airy parts, but they are nevertheless white and take on the luster of fat. However, if I began to set forth an example of proofs especially relevant to anatomy, ✳ I would never do better than go to the vena cava, which Galen argues against Aristotle originates from the liver and not from the heart. In this matter I would not like anyone to think that I have set limits on something, but only that certain reasons of Galen have been examined to prevent anyone imagining a false fabric of the human body because of them.

In the sixth book of *De placitis Hippocratis et Platonis* where the origin of the veins and the principal home of the nourishing faculty are described, several passages occur that are relevant to this topic. When Galen is about to discuss the origin of veins, he applies three proofs to a single axiom, that larger, wider things are the beginnings of smaller, thinner ones.[232] The first proof is taken from plants, whose beginning Galen determines is from their procreation out of a seed. Where they are thickest and where, when the seed swells with moisture and air and breaks out in the earth, we see it sprout something upward and down. When the part that goes upward has grown like a tail for a period of time and then split into many branches, unless the variety of the plant calls for immediate ascent, it is split variously into several parts. Galen therefore determined that the plant's beginning depends upon the point where the roots are seen and the plant is borne upward. That must be conceded to him, since however it is fashioned he derives the middle term of the syllogism in this reasoning from Aristotle rather than himself.

---

[232] For this analogy, see *De placitis Hippocratis et Platonis* 6.3.18.1 ff.: "Just as plants draw all their nourishment from the earth through their roots, so the heart draws air from the lungs through the arteries mentioned. These two very large arteries have grown out of the heart, each from its own shoot. And as in the plant the part that projects from the earth, the stem or trunk, is the widest of all its parts, and the part that divides off into roots is the widest of all the lower parts, and the part between them is the source of the plant, in the same way the largest artery is like the crown of the root for the animal as a whole, and the artery that has its insertion in the lungs is the largest of all therein, while the heart is between the two and the source of the faculties controlling them. In man, being of a more formidable nature beyond what has already been said, it is clear that things greater are sources of things smaller, just as the spring is the source of the streams to which it is distributed. Yet some have been so illogical as to think that what comes after is greater than the source, being misled by rivers that are quite small at their sources and are increased as they move along. But this is not necessarily always the case." (tr. De Lacy 1981, 377–379).

For the same reason we cannot contradict Galen here or in his book *De semine*, where he stated that this was at once the effective and material cause of the fetus and that at one time the substance of the seed is changed into a plant – even if we see that when the new offshoot is separated from the lower part of the seed and strengthened in the manner of roots and the entire seed with its hull is cut apart below, ⁕ it is taken from the ground and soon the divided hull falls away. Then two kernels or parts of the seed grow into a new plant on each side and afterward become green so long as the new plant bears leaves and its kernels or parts of the seed dry up and fall away. The small appendage that stands in the tip of the two parts of the kernel or in certain seeds such as peas and beans is located at about the middle of their side, is seen not to degenerate into the substance of the plant, especially when a seed removed from the ground has already begun to grow and nothing from its form (except for its size) and much less from its substance has been lost.

A second argument from Galen to inform us that the origin of the vena cava does not come from the heart is based upon rivers, which are seen to be greater in their beginning than in the rest of their course; here he warns us at the outset that we should not think of rivers which are increased by the influx of other springs or rivers and easily in their progress surpass the size of their beginning: we should imagine a stream with only a single source that takes on no increment as it goes but scatters into branches and becomes thin like veins passing from their stem.

A third argument is taken with great elegance from other organs in man that remove something from the body. Arteries have a stem coming from the heart, which is then divided variously through the body. The dorsal medulla is also like a stem from which nerves sprout like limbs on a tree.

As soon as Galen has established with these three arguments that greater things are the beginning of lesser, he immediately sums up; but the vena cava is largest and widest in the liver, so he then easily infers that the beginning of the vena cava must be established in the liver, as the place of the vena cava resembles the trunk of a tree, the beginning of a stream, and the origin of arteries; it could also be compared to the dorsal medulla. *

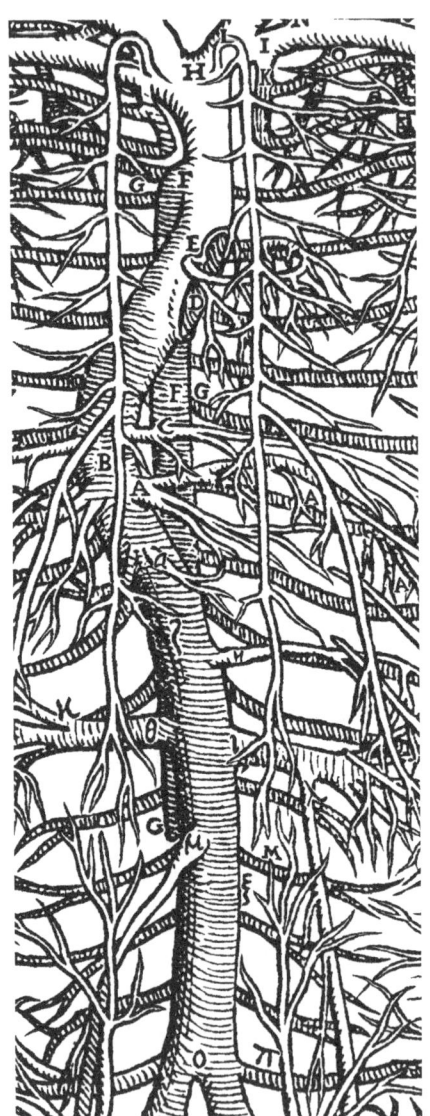

Thus the minor proposition showed Galen nothing relevant, since he believed everything in anatomy could be fashioned as he liked it. He asserted that the vein in the liver is as large as the size of its ascending part and its descending part combined, and wrote that the vena cava is borne straight along the liver, following the opinion of Hippocrates, who was more thorough in this part than he was. Unlike what he said elsewhere, when he attests that its largest stem originates from the hump of the liver in the same way that the great artery proceeds from the base of the heart, and when it is split into two unequal parts, Galen teaches that the vena cava is divided into a greater portion which goes downward and a lesser which is taken upward. After he has described the size of the vena cava

This detail of Vesalius' diagram of the veins preceding the 6th chapter of Fabrica III shows the ascending and descending vena cava. D marks the large orifice of the vena cava opening into the right ventricle of the heart; three A's mark the branch distributing offshoots to the left side of the liver. F, F is the azygos vein, seen originating above the heart.

at the liver, he says that the vena cava reaches the heart as if it were offering it only some branch. When he states in this work that the ascending part of the vena cava above the liver is narrower than the descending so as to add strength to his reasoning there, he teaches that many veins from the vena cava originate between the heart and the liver, chiefly to perform the function of the azygos vein.

Because such statements occur in Galen, much against the liking of philosophers I provide an occasion for them to argue with reasons taken from here that Averroes[233] defended Aristotle in everything for which he is criticized by Galen; I could not fail to warn students while teaching and in my book that the vena cava extends to the heart with a larger orifice than its size in the liver. In fact, this orifice has a capacity equal to the volume of the descending and ascending parts of the vena cava, far from being smaller than the ascending part of the vena cava above the heart, as Galen obviously believes contrary to the truth of the matter. So in the present ✳ body we should, unlike Galen himself, subordinate his lesser propositions to his greater ones according to true anatomy, and then apply them to Aristotle's opinion. It is well known that in this part of the body Galen's arguments are quite ineffectual; we have also shown above that without any reason he makes the part of the vena cava between the heart and the liver narrower than what is below the liver. It is also perfectly clear that the azygos vein and many other offshoots were inaccurately described by Galen in *De placitis Hippocratis et Platonis* as originating beneath the heart.

I am not only astonished at so many false propositions of Galen that he makes to prove his doctrine, and his fictions in anatomy; I am also amazed at the argument that he introduces regarding the course

185

---

[233]   Abu al-Walid Muhammad ibn Ahmad ibn Muhammad ibn Rushd (1126–1198), known in the West as Averroës, worked during the last years of Muslim rule in Spain. In addition to the *Colliget*, a summary of medicine published in Latin in 1527, he wrote *Commentaries on Aristotle*, which was widely admired by Christian Scholastics for its argument that religion is compatible with science and philosophy.

of the vena cava to the base of the heart, stating that if the vena cava originated from the heart like the great artery, it would be split into two trunks so it could conveniently be carried upward and down. This is such bad anatomy that it is scarcely anatomy at all. The great artery originates from the middle of the heart's base and is necessarily taken upward at its stem until it can divide off its largest part. This curves back along the back of the heart to the spine and is distributed to all the parts beneath the heart. It was therefore very opportune and necessary to the artery, which otherwise needed to originate from the middle of the heart's base, to be taken for some time with a single stem no differently than the arterial vein [*truncus pulmonalis*], which because it also comes forth from the middle of the heart's base and not from its side, is borne for some distance with a single stem until it is divided into two trunks. The venous artery [*vena pulmonalis*] is brought out from the left side of the heart's base with a single orifice, but immediately after its beginning it is split into a right and left half so that many anatomists, if only because of Galen's testimony, ascribe two orifices to it, for it would have been quite awkward ✳ for a vessel originating there to be borne for some distance with a single stem because a portion is to be presented to the right lung.

186

Now the vena cava, extending with its orifice to the right side of the heart's base just as the venous artery extends to the left, even if it originated from the heart there, would not have needed to be borne for some distance with a single stem before being split into a part going upward and another that goes down. That is because as the vena cava is now arranged it is taken more or less obliquely along the heart so that its stem will be able to stand in the middle of the body; it would have been crude and without any skill in creation for a single stem to be borne still more to the right and then have its trunks directed to the left towards the middle of the body.[234]

---

[234] The illustration of the inferior vena cava made for the 1543 *Fabrica* and reproduced above does, however, show such a detour to the right and back to the left. By now, it seems, Vesalius realized no such curvature exists.

As I previously warned, Galen attests that the vena cava proceeds with a single stem from the liver and is then divided like the great artery. Yet elsewhere because of Hippocrates' account, which taught that the stem of the vein runs straight up and down, he would appear to have abandoned that view. But however that may be, the stem of the vena cava does not show itself differently in the back of the liver than the way I described it, which is quite different from the way Galen is seen to have described it, whichever of the conflicting passages of his we have judged the truer one. Although he did not write elsewhere that the vein is split in two, he did say everywhere that its orifice in the liver equals the size of both its ascending and descending parts. If he had been committed to dissection rather than his zeal for condemning Aristotle, he would have preserved the utmost truth about the orifice of the vena cava at the heart.

I find nothing weaker than the argument put forward by Galen about the right ventricle or cavity of the heart, which animals without a lung do not have, as if Aristotle, or those who have said the vena cava originates from the heart, had not said that about animals that have this ventricle; ✳ and they wrote no differently about these animals than about those in which there is no right ventricle. Galen should not have made this cavity a cause of objection against Aristotle; it should rather have been explained that the vena cava is provided with one cavity into the heart, not that it sends off an offshoot. I do not think Galen altogether believed that the heart of those animals was completely lacking the offshoot, since their vena cava is connected to the base of the heart just as it is in those animals that have two ventricles. For this reason Galen's logic is not very sound.

When he makes all veins common to the liver and not the heart, what, I beg to ask, is common to the liver with the arterial vein [*truncus pulmonalis*]? Or, if you deny that this is a vein because it has an arterial body, at least the venous artery [*vena pulmonalis*] no less meets the liver than the portal vein meets the heart. What he brings forward about the description or account of the veins is a rhetorical argument.

187

Indeed, if we had not been swearing in this way by the words of Galen,[235] perhaps I would think it more opportune to begin from the heart in my account of the series coming from the vena cava, whose orifice we have said is the largest of all that open into the heart, just as I do for the other vessels of the heart. What would keep us from saying that the vena cava is a double vein immediately upon its origin from the heart, and that with one portion it goes to the upper body and with its other the lower? If the account is begun there where it is largest, we could then add that the part going downward is borne along the back of the liver and extends branches to the liver from its anterior side. Chiefly for this reason the account would appear more fitting, because we cannot describe the makeup of the vena cava from the liver, as I explained at length above. Although one should always begin from the progress of the material which they deliver, since the portal vein provides so much juice to the liver from which blood, both kinds of bile, ✳ and urine result, its description should begin from the intestines: for the blood which we believe flows back into it is in no proportionate measure with that juice. But these are of such a kind that can be carried into both parts. The calculation could be weakened to some degree by the membranes controlling the orifice of the vena cava as compared to the orifice of the venous artery, if one proposes to bring every small factor into the argument.

I would add something about the umbilical vein, if I had examined the vein leading into the mesentery of a canine fetus and into the loins in a human fetus. A calculation could be taken from this and the umbilical arteries to show that it was not necessary for the umbilical vein to be inserted in the heart, since it is more closely shared with the vena cava at the heart than the fetal artery is to the great artery (as even now is always the case).

Because the venous artery does not pulse anywhere in its course like arteries, as the vena cava does not pulse, we are unable to learn

188

---

[235] See n. 190 above about this tag from Horace.

from it that the vena cava does not originate from the heart. To the contrary, we learn that Galen's description can be refuted. At the same time, I do not deny the liver the role of sanguification, which Galen attributed to it at such length. This agrees with Aristotle, though he stated that the spleen is also responsible for sanguification.

I shall say nothing about the first organ in the development of the fetus; for although I could easily reject the opinion of others in many ways, I would introduce nothing that seems sound to me in every way, so obscure to me is everything about the development of the fetus, and full of doubts. Nevertheless I should like to have the reasons available for which Buccaferreo favored Aristotle in the argument about sanguification, since he learned from me the arguments which I have now proposed while dissecting. These relate to something besides the fabric of the parts, and therefore here too I shall not enter into the dispute ✳ where Galen says the stomach is nourished 189 by supplying its own juice which it prepares to the intestines; the liver and all the parts move the material brought to them in such a way that they eventually remove the excess and what is unfit for their nutrition. I believe quite different faculties and actions are present in the parts by which they attend to the nourishment of their own body and by which they serve the body as a whole; I take my argument especially from the uterus and the heart, or even from the bladders and intestines. But because I could not extricate myself here with a short explanation, I shall move along to other topics which it will suffice to note down in three or four words.

An example is the reasoning of Galen in which he affirms that the right kidney is higher than the left because of the directness and ease of attraction.[236] This is not to say that we now and then observe a

---

[236] In *Fabrica* V ch. 10 Vesalius objected to Galen's claim, citing book 5 of *De usu partium*. In his sixth chapter Galen discusses why one kidney is placed higher than another. The kidneys are at different levels, Galen says, because "if they had been placed on a line with one another each would prevent the other from attracting because it would pull in the opposite direction." (3.367.8–11, tr. May 1968, 258).

left kidney higher than the right, but why, I beg to ask, do we determine that directness is different on the right side than on the left,
especially when either side of the stem of the vena cava is equally
embraced by the substance of the liver, and it is known to the student
of true anatomy that the branches presented to the liver are brought
from the front of the vein? You cannot imagine any way in which this
difference could occur in the sides of the vena cava, unless perhaps
one determined to impugn Galen were to propose that the greater
portion of the liver stands to the left of the vena cava than to the right,
taking his argument from the position and course of the vena cava.
The often emphasized demonstration of Galen comes to mind which
he claims is true and irrefutable, deducing with it that the glandular
body positioned at the beginning of the neck of the bladder, admitting the insertion of the vessels delivering semen, does not generate
semen. However, the ancients believed semen is only held by it (as
can be inferred from their words), as we also observe in dissecting that
in no part is there as much semen as in ＊ the glandular body itself.
Galen's demonstration is of this kind. If semen were produced in the
glandular attendants,[237] castrated animals would desire its excretion.
But they are seen not to desire it: therefore, it is quite evident that
semen is not generated in them. How much this demonstration, as it is
not exactly anatomical, falls short of Galen's often repeated and cited
demonstrations rather than those that are consistent with the truth,
will finally be known to anyone who considers that Galen, even if that
glandular body did generate semen, removed the material from which
it is composed. Hence, material suited to be converted into semen

190

The reason for the higher position of the right kidney is that "most of the branches
of the vena cava [*vv. hepaticae*] which bring the blood from the convex part of the
liver open at the right; and it is easier for every body with an attractive faculty to
exert it in a straight line." (3.367.16–368.1, tr. May 1968, 258).

[237] Lat. *glandosis astitibus*, vessels carrying sperm to the prostate gland. Vesalius' term is
derived from Herophilus' παραστάται ἀδενοειδεῖς (glandular assistants), which he
found in Galen's *De usu partium* (4.190.1 ff., tr. May 1968, 644).

could be presented to this body by the seminal vessels which Galen, however, removes by castration, not noticing that a similar demonstration, indeed a completely worthless and poorly constructed proof, can be put forward in this fashion about the testicles. If the testicles acted by generating semen, animals whose vein and artery delivering the material of semen had them tied off where they slip out of the great space of the peritoneum or were cut off in such a way that they remained in the scrotum, would seek the excretion of semen. But they do not seek it. Therefore the testicles do not function to generate semen, in the same way, no doubt, as if testicles were able to prepare semen without material. This is not what I say, as if I were to argue that Galen had wrongly assigned the function of the glandular body, but only so we could ponder the craft of demonstration. We are not unaware how castrated animals sometimes desire venery no less than boys, but in vain.

When I inquired into the reason for the location of the breasts, it did not satisfy me in Galen when he stated from Aristotle's opinion that they are located on the chest because nourishment is taken in other animals into the mane or the horns,[238] ✳ as if quadrupeds lack veins running beneath the breast bone and others through the front of the chest. It is also as if this reasoning were not like the one in which we wrongly assert with the mob that a woman is the only animal except perhaps a mare that is subject to monthly purgations, because in those animals the surpluses are dissolved into horns, feathers, manes, or scales, just as if the stag and the ram did not have longer and larger horns or rather were the only ones of their species so equipped, or the lion were not more hairy than the lioness and the rooster better

191

---

[238] "In animals in which most of the residues in the upper parts of the body are used up in making horns, teeth of a large size, flowing manes, and other structures of the sort, it was natural for another useful residue to be collected in the region of the thorax. Accordingly, in these animals Nature moved the mammae down from the thorax to the abdomen," etc. (Galen *De usu partium* 3.602.7–12, tr. May 1968, 380).

plumed than the hen. But because such matters have more to do with Aristotle's problems[239] than with Galen's anatomy, we should pass them over.

Besides the fact that I should have liked Galen to have a better knowledge of the interval between the veins that go to the uterus and to the lower abdomen, he does not seem to me to have a correct understanding when he takes blood to the breasts from the veins coterminous with the uterus, because there are no veins into which they could more readily unburden themselves. It should have been known to him how tiny those veins are which run upward and how thin, how great the size of the vena cava, and how direct its descent into the veins of the legs. There is also the swelling of the abdomen, which prevents the upward passage of blood; so far am I from wishing to state with Galen that the swelling and tightening of the abdomen aids the ascent of blood. These have more to do with probable and rhetorical propositions than with anatomy, against which should be tested so many vivisections[240] that are misleading because they show the role of none of the intercostal muscles in opening the thorax or drawing the ribs together. If I were to undertake to explain these vivisections, this piece of writing would grow too far beyond my intention. However, ✳ as an example I shall at least recall one; no other vivisection in Galen is better known than the one used in *De usu pulsuum, De placitis Hippocratis et Platonis*, and elsewhere, where he says that everything in the neck must be cut away by which the heart can communicate with the brain, whether it travels up from the heart or from the brain down. But if we think about this vivisection carefully, two arteries present themselves to us which empty into the first sinuses of the hard membrane, and then

192

---

[239] Not the pseudo-Aristotelian *Problemata* but *De generatione animalium* 728b18 ff.: "... the human body does not possess the sort of parts to which the [menstrual] residue gets diverted, as other animals do: it has no great coat of hair all over the body, and no secretions in the form of bones, horns, and tusks." (Loeb tr. by A. L. Peck, p. 107).

[240] Treated at length in the 19th chapter of *Fabrica* Bk. 7: "Some Remarks on Vivisection."

from the series of cerebral vessels to the network [*plexus choroideus*] compared to the secundines [*placenta*] or outer wrapping of the fetus; these arteries deliver material from the heart, and in his cutting he separates them intact and whole. But many such errors occur elsewhere, which students of anatomy must examine closely if they wish to declare themselves Galenists (as we surely all should). For nothing in all the books of Galen seems difficult to me except in his anatomical works, in which Galen is seen to be quite assiduous. So when we are trained in the first disciplines we find few matters which will delay us in the reading of Galen except anatomy.

How useful the annotations of Vesalius have been in Galenic anatomy, and how little they are to be needed hereafter

I first began to put this to the test while preparing annotations on the anatomical works of Galen, for example from his inconsistent statements and variant descriptions of things, many of which I have mentioned in this epistle; many descriptions are also considered having to do with the observation of names which he himself put in place. Likewise I have remarked upon differing kinds of bone structure, the enumeration of bones of the upper maxilla, the vertebra taken up on both sides,[241] the sacrum, the processes of the scapula, and primarily their names; the description of the humerus along its length; the numbering of the foot bones; the muscles of the eyelids, and of those that move the lower jaw; about the muscles of the hyoid bone; about the muscles moving the scapula, the forearm, ✳ the wrist, and the fingers; about the eight that are ascribed to the abdomen; and about those that control motions of the thigh, lower leg, and toes. It would be easy enough to recount various opinions, and from that it would readily be known what Galen added since the first publication of *De*

193

---

[241] The twelfth thoracic vertebra, whose structure is uniquely intermediate between those above and those below. Galen, who relied upon animal specimens, had variously said this was the tenth or the eleventh.

usu partium and when he made use of the writings of other anatomists rather than his own inspection of the matter, which Sylvius is seen to resent that I too have pointed out. If, I say, I have written an unjust book in undertaking this epistle to you, which betokens my friendship toward you and the pleasure I take in casual conversation and discussion with you: whatever this is, I shall have written it with little trouble. I take a supreme pleasure in the memory of those things which I have learned with such drudgery and squalor, in return for which I deserve something far different from the calumnies of those who because this exertion did not turn out as well for them and the subject does not yield them a ready reputation, have spewed their bile upon me, being so noxiously irate with me because I do not agree with Galen and therefore do not withhold all faith from my eyes and my reason.

It is interesting to me that as soon as Sylvius seemed to become a little milder in his letter, he added that he very much wished I had presented my argument more gently and in a different manner; here he had the suspicion that not everything in Galen was sound, and he therefore wrote that he wanted the things I believed need annotation in Galen to be added to his books, in the same way as Sylvius' disciples annotate from my book, the ones he hopes to persuade to sharpen their pens against me (though he is the one who does so, and abuses the name of disciple). It is as if I could institute something better than a new description of the entire human body when everybody has fallen so short of it and Galen is so scanty in descriptions of things though he has written the densest books. I have yet to see how I could better have ⁎ presented my studious labors so that everything could turn out as desired at absolutely no expense and spare enough to bring something into a common summary of my endeavors proportional to the slenderness of my talent.

Sylvius urges that I distil my annotations of Galen well and at length (which I remembered as opportunity provided while reconstructing many passages in the first book of *De anatomicis administrationibus*, in my chapter about the muscles moving the forearm).[242] He assumes that because of my youth and hasty publication I have made some error in my book which I now regret, and that my present time of life, not free of a wife, children, or any care of private business, in the delightful company of friends, would be perfectly suited to writing about anatomy. It was, in fact, no pleasure for me to turn over bones for hours on end at Paris in the Cemetery of the Innocents, or to go to look at bones at Montfaucon,[243] where once with a companion I was nearly killed by so many murderous dogs, or let myself be locked out of Louvain by myself so I could take bones from a gibbet in the middle of the night to prepare a skeleton. I will no longer be a nuisance to judges to have people killed with one form of execution or another or reserve them for this or that opportune time for our dissections. I will not keep warning all my students to be on the lookout when someone is to be placed in a tomb, nor will I now urge them to observe in what condition someone died under the care of teachers whom they attended to learn their craft (for which a welcome reward would eventually come to all). I will not store up in my room for weeks on end bodies taken from graves or given to public execution, nor will I surrender myself to sculptors or painters to be so vexed that often because of their peevishness I thought myself more hapless than the corpses who had fallen to my lot for dissection. All of these things * my immature age easily and cheerfully endured 195* for the development of my craft and my passion to learn and advance our common interests.

I shall pass over the hard work with which I taxed myself for at least three full weeks when those who were present while I

---

[242] Chapter 46 of Bk. II of the *Fabrica* refers more than once to defects in Galen's description of these muscles in *De anat. adm.* In the 1555 edition, he promises these will be more fully explained "in my annotations to the Anatomical Works of Galen."

[243] Site of the principal gallows controlled by the kings of France, northeast of Paris.

* In the 1546 edition, page 195 was misnumbered as 196.

taught anatomy in Italy saw me in public dissections, though I once performed this task at three universities in a single year.[244] So if I put off writing until the present time, and now for the first time am undertaking a condensed version,[245] my anatomical efforts would not be in the hands of students; I do not know whether works will be presented by later generations from a boiled-down Mesue,[246] a boiled-down Gatinaria,[247] or a boiled-down *De morborum differentiis* of some Stephanus[248] and *De causis morborum, De symptomatum differentiis, De symptomatum causis*, a book about tables, or finally a boiled-down part of Servitor's drug sellers.[249]

So far as concerns my Annotations which had grown into a huge volume, those along with a complete paraphrase of the ten books of Rhazes' *To King Almansor*[250] which I wrote much more carefully than

---

[244] Following the Geldric War that ended in September 1543 (see n. 223 above), Vesalius lectured at Padua, Pisa, and Bologna.

[245] A *coctio* or boiled-down version of the *Fabrica*'s critique of Galen corresponding to the condensed versions of Mesue, Gatinaria, and other authors for medical students.

[246] Mesue the Younger or Johannes son of Mesuë (Masawaih al-Mardini), 10th-century Arabic author of *Canones Generales*, the subject of a commentary by Mondino de' Luzzi (ca. 1270–1326). He was also valued in the West for his textbook on pharmacy.

[247] Marco Gatinaria (1442–1496), author of *De curis egritudinum particularium noni Almansoris practica uberrima*, published in 31 editions in three languages between 1506 and 1539. Another work, *Morborum internorum prope omnium curatio, brevi methodo comprehensa*, appeared in five editions between 1545 and 1549.

[248] Stephanus of Athens (c. 550–622), the Byzantine philosopher and physician, wrote commentaries on Hippocrates' Aphorisms, Galen's Therapeutics, Galen's *De morborum differentiis*, and a book on uroscopy.

[249] Abu al-Qa-sim al-Zahrawi (936–1013), known in the West as Albucasis, wrote a book on the preparation of medicines which was translated into Latin in 1288 by Simon of Genoa, *Liber Servitoris de praeparatione medicinarum simplicium* printed at Venice in 1471. "It was relied on by many generations of European apothecaries as a first-hand source of information on pharmaceutical processes." (Sneader 2005, 26).

[250] Rhazes' *Liber ad Almansorem*, on which see n. 188 above. Vesalius' continuing interest in this medieval Persian author (865–925) is one of several habits of mind that distinguish him from the Humanists who gave him his early training.

my paraphrase on the ninth book which is in print, and the beginnings of a book on the formulation of medications (into which I had gathered many medicines that I thought useful) – all these perished in a single day with all the books of Galen that I had used in learning anatomy and had stained in various ways, as usually happens. When I left Italy to enter the court, and the doctors whom you know made the worst criticisms of my book and of all the books that are published today to advance study, addressing the Emperor and other powerful men, I burned everything (intending in the future to have little difficulty refraining from writing), though more than once I regretted my petulance and was sorry not to have stopped myself at the warnings of friends who were present.

But concerning my Annotations I am for that reason quite happy * that no desire to publish them is likely to come over me (even if they were extant), since I could easily predict how hostile they would have made everyone to me even if so few matters at odds with the views of Galen have chanced to make their way into my book. They have aroused anger in so many critics and surrounded themselves with so many defenders of Galen after three years, and even before the publication of my book. In a word, I believe that having been prepared after so long an interval of time they will make common cause with students and will come out with more than a short epistle full of accusations and without explication of all the passages in Galen.

Concerning the books of Galen I have nothing to lament, since they have chanced to fall into the hands of people who lacked the ability to distinguish bad things written in the margins from the good. I believe you know how many marginalia are made in the universities by teachers; as soon as we read something we write things in the margins that later seem to us inept and ridiculous. For my part I can make the guess that I read out Galen's *De ossibus* to students as many as three times before I dared to note any correction to Galen, while now I cannot sufficiently wonder at my stupidity that I so badly understood what was written and that I so cheated my own eyes.

Because I had the greatest pleasure in making a paraphrase comparing the Arabs with Galen and other Greeks in the parts of medicine about which Rhazes wrote in each of the books of his treatise, I am grieved that this project is lost to me, especially for the sake of my grandfather Everard,[251] whose commentary on that work of Rhazes I believe is not without learning; his commentary on the first four sections of Hippocrates' *Aphorisms*, and several mathematical works attest that he was a man of singular talent. I have heard more than once from my father,[252] * of pious memory, that though Everard had begun to take the post of my great-grandfather Johannes[253] as physician to Mary, wife of the emperor Maximilian, he passed away before he was 36 years old. Johannes survived him by many years and taught medicine at Louvain at an advanced age that was not suited for court life. That his father Peter[254] was a doctor is known from his book on the fourth Fen of Avicenna and several books inscribed with his name that are found among volumes that my mother still saves. Among them we have all the books that were copied out at huge (but now useless) expense, which were familiar to the doctors of that time. The record of these parents and their memory was quite

---

[251] Everard de Wesalia (d. ca. 1485), physician to the Archduke Maximilian. See Spelkens 1961, 67 and O'Malley 1964, 25 f., who doubts Vesalius' report here of his youthful death.

[252] Andries Van Wesele (b. ca. 1479), the illegitimate son of Everard Van Vesele and Marguerite s'Winters. He served as apothecary to Margaret of Austria and later Charles V.

[253] Johannes de Wesalia, born soon after the beginning of the 15th century (d. 28 May 1476), father of Everard. The wife of Maximilian was Mary of Burgundy, whose son Philip was to be the father of Charles V.

[254] Peter Witing bore the ancestral surname before his son Johannes adopted the name of their native town, Wesel in Cleves, as his surname. O'Malley 1964, 21 describes the family history given here as "a mixture of fact and legend." The five generations of his family recorded here by Vesalius may be summarized (1) Peter Witing; (2) Johannes de Wesalia; (3) Everard de Wesalia; (4) Andries Van Wesele; (5) Andreas Vesalius.

pleasant and revered when I was at Nymwegen; at the ancient and celebrated Wesalia in Cleves where they originated I was able to see the tombs of the Witing family when because of the adverse health of B. Navagero[255] it was necessary to delay at Nymwegen so long after the departure of the Emperor.

So far as concerns me, it was at no one's urging that I had the task of boiling down my Annotations to Galen, though it can be inferred from what I have said here in response to Sylvius that they would have been by no means useless. They may perhaps have been exposed to condemnation because they would have revealed the negligence in anatomy of many who have claimed a great name for themselves in that field. In addition, I had showed in my annotations that a third type of disease, centered in a continuous loosening,[256] was correctly reported by the ancients, differently from Galen, and that Galen faulted them undeservedly in this respect, as he often did in other ways. It was also shown that the ninth temperament was taken from Chrysippus, though Galen's authority placed it elsewhere.[257] Certain divisions in medical art were also shown not to match the method of his own art. ✳   198✳

---

[255] On whom see n. 11 above. O'Malley (1964, 21) understands this to mean that during his delay at Nymwegen Vesalius was able to make a side trip to Wesel.

[256] One of the three general bodily conditions (dryness, fluidity, and a mixed condition) described by the Methodist sect, which Galen strenuously condemned. In his portrait used as the frontispiece of the *Fabrica*, the *Epitome*, and the China Root Epistle, the words OCYUS IUCUNDE ET TUTE, a variant of the Methodist motto TUTO CELERITER IUCUNDE, are shown carved on the table at which Vesalius is standing. The Methodist ideal was that every procedure should be "safe, swift, and pleasant" (viz. painless).

[257] Galen's *De temperamentis* made the ninth temperament a steady mixture of warm, moist, cold, and dry. Chrysippus of Soli (ca 279– ca. 206 BCE), the third head of the Stoic school of philosophy, wrote a treatise on psychology *On the Soul*, now lost except for fragments, which Galen criticized.

✳   In the 1546 edition, page 198 was misnumbered as 199.

There were besides other facts which this is not the place to recount, that prove it is not my habit to weave new garlands for myself from the boiled-down labors of others, or to bring forth works written by others dressed up with some splendor or another of diction or paraphrased (not to say made more obscure) as if they were my own. And what desire, I ask, could anyone have for the publication of his own lucubrations when there is nowhere a lack of people plotting to destroy the labors of others? Even England now shows where they have counterfeited the figures from my *Epitome* so obscurely and without skill in the engraving though not without expense: yet someone brought it out in such a way that it would embarrass someone to believe it had been published that way by me. To take a single example, the course of vessels, which my friends know I drew myself in my book, was so distorted that I can see scarcely any evidence of my diligence. Moreover, everything in England[258] is hideously reduced out of proportion, though images of this kind can never be printed large enough. I wonder why those grossly incompetent imitators have not observed that while I publish a register of letters of privilege or rights granted to me I would much rather grant publishers permission to print my figures than have them badly imitated by others. Though I gladly wrote that at the time, I would now like to have the same thing said to all because I would prefer to suffer the greatest sacrifice of my private property and would more willingly furnish something for the elegance of an edition than to have things produced for me by my uncommon labor so hideously marred. Everyone will believe good editions so much better than to have new editions produced with more inflated titles and more authors.

So may the Gods support my work, I must overlook no means by which I can otherwise resist those imitators ✳ and the people who lie in ambush against the efforts of others when they can clear their own nostrils of nothing new.

---

[258]   See n. 4 above on the 1545 Geminus plagiarism published in London.

In the meantime, be in good health and commend me to our common friends; have my letter together with yours given to them, as you will be glad to present them (if they ask it of you) with whatever labor you have spent in writing to me.

Ratisbon, written the Ides of June, year of salvation 1546.

Your most devoted

Andreas Vesalius

An Italian Treatise
in which is investigated

THE METHOD OF ADMINISTERING THE CHINA,

added here as it was sent
BY ANDREAS VESALIUS TO JOACHIM ROELANTS

FTEN it happens, my most learned Joachim, when I explain in a letter to you the method of administering and preparing the decoction of the China root, that mention is made of an Italian treatise which I am giving to you written out exactly as it was sent to our court by others, so you will know how much has come into my hands about this new and already famous medicament. Though thanks to my friends I have given it in Latin, I send it in Italian (as I received it) because I know you are not ignorant of this tongue. You will also by this means more exactly gauge the mind of the author (whoever this empiric was) and feel I have satisfied your request. I have little doubt, however, that it was translated from the Spanish, since I also have at hand a Spanish formula together with a method of boiling down and administering the trifling Sparta parilla. *

201 But the Spanish description is much shorter than the present Italian and more fragmentary, alien to the method of the art in other places.

REGIMENTO PER PIGLIAR L'ACQUA DE LA RADICE DE CHYNA

Method of Administering
the Water of the China Root
[*Translated by Francesca Tataranni*]

A purgation will be carried out at the beginning, middle, and end on the advice of the physician, who will take into account the condition of the patient. Twenty-four ounces of the root should be divided into twenty-four parts to make a fresh decoction every day. That which you will put to boil tomorrow, on the previous day cut into small pieces, the smallest possible, and put them in a little water, leaving them to soak until the next day. Then put this root and the infusion into a new pot and pour three pitchers of spring water into it; let it boil until a third has evaporated. The pot should be of a size to ensure that the boiling decoction does not overflow; it should always be covered so its potency does not escape. Once cooked and removed from the fire, it needs to be covered with large towels lest it cool off completely, and it needs to be made fresh every day because otherwise it would become acid. If the patient is unable to have physical benefit, add to each preparation of this water half of an eighth ounce of celery root cooked with this China root.

202 The patient will take a large glass of this decoction hot * early in the morning while staying in bed, and thereafter will remain covered

in bed and try to sweat for the space of two hours: the more he sweats, the more benefit he will have. After becoming dry, he should get up and walk, keeping himself out of the wind and well covered with clothing, especially for seven or eight days, during which the body should be in motion. After eight days have passed, if he likes he will be able to go outside, keeping himself out of the wind and wearing warm clothes. Throughout the day when he wants to drink he should drink the aforementioned decoction, which needs to be tepid, the warmer the better. He should eat young poultry or capons boiled without salt. He must avoid vinegar and anything acidic for fourteen days; he will eat nothing roasted. He will be able to eat all kinds of preserves and quince jams made with sweetening; he will not eat dairy products, and he will eat quince jams after meals or other things that benefit the body, especially honey, which is the best thing one can eat and can be eaten at any time. Everything should be eaten in moderation, because it is judged that diet has as much benefit and effect as the decoction. One should not drink wine, broth, or anything else during the twenty-four days that the aforementioned root lasts, and for this time one needs to refrain from sex. Once all the water is cooked and the root administered, the small pieces should be set to dry in the sun or elsewhere where they can dry, and once they are dry they are cut into smaller pieces. Two ounces of these pieces are to be placed in the amount of water previously mentioned and made to cook as prescribed above. This water should be drunk another eight or ten days ✳ in addition to the twenty-four. If anyone who takes this decoction should have sores, he should place upon them nothing but towels soaked in the decoction and stay at home. Anytime he wants to go out he should put on some others that are not soaked and upon returning put the soaked towels back on. He will often bathe the sores with the decoction, which has been so effective on sores that it could be no better.

He should not eat fish of any sort, though sometimes good fish is not forbidden. Thus if the patient is weary he may drink wine diluted with the decoction, and they let him eat any food that gives nourishment. They generally forbid all the following things, such as women, vinegar, salt, ingredients containing acids, herbs, goat meat or he-goat, and rooster: all these things except roast meat, which they deny him for fifteen days. From seven days before then he will begin to feel great pain in the parts where he is sick, and the pain will go on increasing until the fourteenth or fifteenth day. Thereupon he will feel well, and the sores will be healed (God willing) because the virtue of this root is great, and the patient should make every effort to complete the treatment and drink the decoction very hot in the morning and tepid the rest of the time. He should make himself sweat, because the more he sweats the better he will feel, and the greater will he will have to eat. The more days he takes the decoction at the weight and measure stated above, the healthier he will stay and the more it will profit his body; it will purge the inner parts without making him spit: not through benefit of the body but through the special virtue that this root has. After seven days he will scarcely have * any physical benefits, because the body must be aided with clysters, not those of pharmacists or their products, but only those made from chicory water or from borage, with rose oil, or in fact common oil with salt. He will follow this regimen for six or seven days. If it is seen that he has physical benefit, he will not leave off going outdoors but he will keep himself out of the wind and keep well clothed, as we have stated above.

PROSOPOGRAPHY OF EARLY MODERN PERSONS MENTIONED IN THE CHINA ROOT LETTER

(Page numbers are those of the 1546 edition.)

Albius, friend of Vesalius at Pisa, student of anatomy, p. 176.

Marcantonio Belloarmato, jurist of Siena: pp. 173, 175.

Buccaferreo, friend and student of Vesalius at Pisa, pp. 176, 188.

John Caius (1510–1573), accused of publishing an inferior expanded version of the *Epitome*: Preface p. 4.

Francesco Campana, private secretary to Duke Cosimo: Preface p. 4; 175.

Stephanus de Casala (named in Index as Stephanus Sala), surgeon to Charles V: p. 36.[1]

D. Cavalius or Caballus, one-time physician to Charles V: pp. 3, 12, 16. See Roth 1892, 203 n. 3.

D. Marzilius Colla, master of the Imperial horses: p. 36.

Realdo Colombo, the unnamed "dabbler" on the medical faculty at Padua who believed he had discovered the vessel conveying black bile from the spleen to the stomach: p. 136.

---

[1] Huard and Imbault-Huard 1980, 15 lists Charles V's medical staff as Cornelius Van Baersdorp, Vesalius, Petrus Lopez, Jacobus Olivarius, Gregorius Lopez, Gonzales Muñoz, Simon Guadalupe, Stephanus de Bourgogne, and later Henri Ma Thys (1500–1563).

Janus Cornarius (c. 1500–1558), in 1546 was professor of medicine
at Marburg, returning in that year to Zwickau. "Primarily a
medical philologist who edited a number of classical medical
texts including works of Hippocrates, Galen, Aetius, and Paul of
Aegina and belonged to that group of physicians who, because
of their strong and constant belief in the validity of classical
authority, were naturally unsympathetic to Vesalius's purpose."
O'Malley 1964, 456 n. 178: p. 178 *bis*, 179.

Cornelius von Baersdorp, second in rank of the Emperor's physicians
pp. 12, 16.

Cosimo de' Medici (1519–1574), Duke of Florence from 1537 to 1574,
reigning Grand Duke of Tuscany from 1569, who offered Vesalius
a post on the medical faculty at Pisa: Preface pp. 1, 5; pp. 40, 140.

D. de Bossu Jean de Hénin-Liétard, (aka le Grand, 1499–1562): childhood friend and Grand Equerry of Emperor Charles V, who
made him Count de Boussu in 1555. Scion of an aristocratic
family in Hainaut, Belgium; his castle was visited by Charles V in
1545 and 1554. pp. 17, 26.

Cardinal Doria: Girolamo Doria (1495–1558), son of Admiral Andrea
Doria (1466–1560): p. 37.

Johannes Dryander (1500–1560), professor of mathematics and medicine at Marburg from 1535, author of *Anatomia capitis humani*
(1536): pp. 177, 178.

Ioannes Eck, distinguished physician at Cologne, common friend of
Johannes Dryander and Vesalius: p. 177, 178.

Anna van Egmont (1533–1588), referred to as *Comitissa Egmondana*:
p. 141.

Jean Fernel (1497–1558), who first used the word "physiology" to
describe the study of bodily function (see Sherrington 1946).
Recommended as a professor at Paris for the son of Joachim
Roelants: p. 42.

Leonhart Fuchs (1501–1556), humanist physician and botanist,
author of the great illustrated herbal *New Kreüterbuch* (1543). An

admirer of Vesalius, in 1551 he wrote an epitome of Vesalius and Galen: p. 179.

Jean Baptiste Gastaldo, a prominent patient who benefited from treatment with the China root: p. 14.

Ioannes Guinter von Andernach (Johann Winter, 1505–1574), Vesalius' professor at Paris, 1534–1536. Translator into Latin of Galen's *De anatomicis administrationibus* (9 vols., Paris 1531); author of *Institutiones anatomicae* (4 vols., Paris 1536), a standard work for physicians published with Vesalius' emendations in 1538: p. 177 *bis*.

Lord of Hallewyn, casuality of the 1544 Battle of Saint-Dizier: p. 176.

Prospero Martello, Florentine patrician whose death was investigated by Vesalius: p. 175.

Giovanni de' Medici or Giovanni dalle Bande Nere, 1498–1526, father of Cosimo de' Medici, famed for his exploits as a condottiero or mercenary military captain: Preface p. 7.

Bernardo Navagero (1507–1565), Venetian ambassador to the imperial court from 1543, overseer of the university at Padua and friend of Vesalius. His illness at Nymwegen in 1546 was the occasion for Vesalius' delay there until April 11, during which he wrote the *Epistle on the China Root*: pp. 11, 197.

Oliveri, member of the faculty of medicine at Paris: p. 42.

Joannes Oporinus (Iohann Herbst), printer of the *Fabrica* in Basel: Preface p. 4.

Prince of Orange: see Hallewyn.

Ludovico Panizza, physician in Mantua and medical author: p. 37.

Juan Bautista Recasulano, bishop of Cortona and ambassador from the duke of Florence to Charles V 1543–1545: Preface p. 6.

Joachim Roelants (1495–1558), protomedicus at Mechlin, author of letter to Vesalius requesting information about the China root. An old and close friend of Vesalius; his son of the same name had gone to Paris to study medicine: pp. 11, 200.

Ludwig Sanches, protonotary of the king of Sicily; pp. 17, 29.

Jacobus Scepperus, copier of the China Root Epistle, who delivered a copy to Franciscus Vesalius: Preface pp. 3, 4.

Jacobus Sylvius (Jacques Dubois, 1478–1555) lectured on medicine at the Collège de Tréguier, Paris, outside the faculty of medicine after receiving his baccalaureate from the Paris faculty of medicine in 1531. After 27 January 1536, he became eligible to lecture for credit from the faculty of medicine, but was not a member of that faculty. Pp. 9, 41–46, 75, 95, 122, 127, 148–149, 151–155, 176–177, 193–194, 197.

Jean Vasses of Meaux, dean and chief administrative officer of the faculty of medicine at Paris (see O'Malley 1964, 37f.): p. 42.

Franciscus or François Vesalius, younger brother of the anatomist and author of the preface to the China Root Epistle: Preface, p. 3.

Gerard Vueldwick, medical colleague of Vesalius, expert on botanicals and leader of a legation to Turkey: pp. 19, 37–39.

Antonio Zuccha, friend of Vesalius: p. 12.

ℰ⌒⊃ BIBLIOGRAPHY

Arrizabalaga, Jon, Henderson, John, and French, Roger. 1997. *The Great Pox. The French Disease in Renaissance Europe.* New Haven: Yale University Press.

Bensky, Dan, and Gamble, Andrew. 1986. *Chinese Herbal Medicine. Materia Medica.* Seattle: Eastland Press.

Bradford, William. 1850. *Correspondence of the Emperor Charles V and his Ambassadors at the Courts of England and France.* London: R. Bentley.

Brain, Peter. 1986. *Galen on Bloodletting: A Study of the Origins, Development, and Validity of his Opinions, with a Translation of the Three Works.* Cambridge and New York: Cambridge University Press.

Brandi, Karl. 1939. *The Emperor Charles V.* Translated from the German by C. V. Wedgwood. New York: Alfred A. Knopf.

Choulant, Johann Ludwig. [1920]. *History and bibliography of anatomic illustration in its relation to anatomic science and the graphic arts.* Chicago: The University of Chicago Press.

Cushing, Harvey W. 1962, 1986. *A Bio-Bibliography of Andreas Vesalius,* 2nd edn. Hamden, CT: Archon Books. Re-issued with new introduction Winchester: St. Paul's Bibliographies.

De Lacy, Phillip. 1980–81. *Galen on the Doctrines of Hippocrates and Plato.* Edition, Translation, and Commentary. First Part: Books I-V, Second Edition; Second Part: Books VI–IX. Berlin: Akademie-Verlag.

1992. *Galen On Semen.* Edition, translation, and commentary. *Corpus Medicorum Graecorum* 5.3.1. Berlin: Akademie Verlag.

Eriksson, Ruben, ed. & tr. 1959. *Andreas Vesalius' First Public Anatomy at Bologna, 1540. An Eyewitness Report by Baldasar Heseler.* Uppsala and Stockholm: Almqvist & Wiksells.

Garrison, Daniel H., and Hast, Malcolm H. 2013. *The Fabric of the Human Body* (annotated translation of Andreas Vesalius, *De humani corporis fabrica* (1543, 1555). Basel: Karger Publishing.

Goss, Charles Mayo. 1961. "On Anatomy of Veins and Arteries" (tr. of Galen *De venarum arteriarumque dissectione*). *The Anatominal Record* CXLI, 355–66.

Hankinson, R. J. 2008. *The Cambridge Companion to Galen.* Cambridge, England: Cambridge University Press.

Huard, Pierre, and Imbault-Huart, Marie-José. 1980. *André Vésale iconographie anatomique (Fabrica, Epitome, Tabulae sex).* Paris: Les Éditions Roger Dacosta.

Kusukawa, Sachiko. 2012. *Picturing the Book of Nature. Image, Text, and Argument in Sixteenth-Century Human Anatomy and Medical Botany.* Chicago: University of Chicago Press.

Lind, L. R. 1975. *Studies in Pre-Vesalian Anatomy. Biography, Translations, Documents.* Philadelphia: American Philosophical Society.

May, Margaret Tallmadge (tr.). 1968. *Galen on the Usefulness of the Parts of the Body.* 2 vols. Ithaca, NY: Cornell University Press.

O'Malley, Charles D. 1958. "Some Episodes in the Medical History of Emperor Charles V." *Journal of the History of Medicine and Allied Sciences* XIII (4), 469–82.

1964. *Andreas Vesalius of Brussels 1514–1564.* Berkeley and Los Angeles: University of California Press.

O'Malley, Charles D., and Saunders, J. B. de C. M. 1950. *The Illustrations from The Works of Andreas Vesalius of Brussels.* World Publishing Company. rp. 1973 Dover Publications.

O'Malley, Charles D., and Moes, R. J. 1960. "Realdo Colombo: 'On Those Things Rarely Found in Anatomy.'" *Bulletin of the History of Medicine* 34: 508–28.

Quétel, Claude. 1990. *The History of Syphilis.* Baltimore, MD: The Johns Hopkins University Press.

Roth, Moritz. 1892, 1965. *Andreas Vesalius Bruxellensis.* Berlin: G. Reimer; repr. Amsterdam: Asher & Co.

Saunders, J. B. de C. M., and O'Malley, Charles D. 1947. *Andreas Vesalius Bruxellensis: The Bloodletting Letter of 1539. An Annotated translation and Study of the Evolution of Vesalius' Scientific Development.* New York: Schuman.

1950. *The Illustrations from the Works of Andreas Vesalius of Brussels.* (Rp. 1973 New York: Dover Publications) Cleveland, OH: World Publishing.

Schmitz, Rudolf and Tan, Freddy Tek Tiong. 1967. "Die Radix Chinae in der 'Epistola de radicis Chinae usu' des Andreas Vesalius (1546)." *Sudhoffs Archiv* 51.3: 217–228.

Sherrington, Charles Scott. 1946. *The Endeavour of Jean Fernel.* Cambridge, England: Cambridge University Press.

Shu, X. S., Gao, Z. H., Yang, X. L., 2006. "Anti-inflammatory and anti-nociceptive activities of Smilax china L. aqueous extract." *Journal of Ethnopharmacology.* 103(3):327–32.

Singer, Charles Joseph. 1952. *De ossibus ad tirones,* tr. by Charles Singer, "Galen's Elementary Course on Bones," *Proceedings of the Royal Society of Medicine* 45, 25–34.

1956. *Galen on Anatomical Procedures.* Oxford University Press.

Siraisi, Nancy G. 1987. *Avicenna in Renaissance Italy.* Princeton University Press.

1990. "Giovanni Argenterio and Sixteenth-Century Medical Innovation: Between Princely Patronage and Academic Controversy" *Osiris* 6:161–80.

Smith, Bonnie J. 1999. *Nvms Canine Anatomy*. Lippincott Williams & Wilkins.

Smith, Wesley D. 1990. "Pleuritis in the Hippocratic Corpus, and After." *La maladie et les malades dans la Collection hippocratique.* Actes du VIe colloque international hippocratique (28 septembre–3 octobre 1987, Québec), ed. Paul Potter, Gilles Maloney, Jacques Desautels. Québec: du Sphinx, 189–207.

Sneader, Walter. 2005. *Drug Discovery. A History*. Hoboken, NJ: Wiley-Interscience.

Spelkins, Emile. 1961. Généalogie de la famille d'André Vésale (Wijtinck dictus van Wesele)" *L'Intermédiaire des Généalogistes* 16:65–67. Brussels.

Tyler, Royall. 1956. *The Emperor Charles the Fifth*. Fair Lawn, N.J.: Essential Books.

van Male, Guillaume. 1843. *Lettres sur la vie intérieur de l'empereur Charles-Quint, écrites par Guillaume van Male.* ed. Frederic Auguste F. T. Reiffenberg. Brussels: Delevigne et Callewaer..

von Staden, Heinrich. 1989. *Herophilus: the Art of Medicine in Early Alexandria*. Cambridge, England: Cambridge University Press.

$\mathcal{C}\!\!\sim\!\!$ VESALIUS' INDEX OF WORDS AND SUBJECTS

(Page numbers in parentheses are those of the 1546 edition.)

247